COLOUR AND VISION
Through the eyes of nature

STEVE PARKER

Published by the Natural History Museum, London

First published by the Natural History Museum, Cromwell Road, London SW7 5BD
© The Trustees of the Natural History Museum, London 2016.

ISBN 978 0 565 09389 1

10 9 8 7 6 5 4 3 2 1

A catalogue record for this book is available from the British Library.

Internal design by Mercer Design, London
Reproduction by Saxon Digital Services
Printed in China by Toppan Leefung Printing Co., Ltd.

Contents

Introduction

I t's estimated that three-quarters or more of the information in the human brain – facts, scenes, words, events, faces, places – comes through the eyes. Within the animal kingdom, mammalian eyes are fine examples of this sense organ, and the human eye in particular is a good example partly due to our evolutionary heritage as members of the group Primates. Other primates, and especially our closest living animal cousins, apes and monkeys, also lead heavily vision-based lives. We literally see the world from a human perspective.

Light, vision and colour dominate the living world. More than 90 per cent of all animal species have image-forming eyes or visual processing of some kind. So vision must confer an immense evolutionary benefit. The advantages of a particular kind of eye depend on its time, place and owner, and they act by the eye's ability to guide or direct behaviours and survival tasks, such as locating food and finding shelter. Over millions of years the growing complexity of visual systems, from basic eyespots to the immensely sophisticated anatomy and workings of eyes like ours, have been associated with increasingly advanced vision-based behaviour.

Like any other body parts, eyes have developed through the process of evolution by natural selection. The evidence for this is all around us. It includes: the wide-ranging variety of eye designs belonging to the vast array of creatures alive today; the instructions, or genes, within living organisms for making eyes;

OPPOSITE Flying insects depend greatly on vision to navigate, locate food and mates, and avoid danger. The golden-eyed horsefly, *Chrysops quadratus*, has eyes many times the size of its brain.

other building-block materials determined by these genes; and the fossilized eyes of long-gone animals preserved in the rocks. The overall picture is that many different kinds of animal eyes have evolved. The most complex and intricate of these developed through a sequence of stages from simpler, more modest predecessors, while remaining useful to their owners at each stage.

Eyes are one part of the vision story. Light must shine onto and into them, so they can detect colours, shapes and patterns, and send nerve signals to the brain. Here, in the brain's vision centres, the signals are decoded, analysed and assembled into an image. This is what we perceive in our conscious awareness – our 'mind's eye'. We have first-hand experience of how humans do this. But do other animals, some with very different eyes and brains, see the world as we do?

One of light's most luminous qualities is colour. The energy of light can be imagined travelling as undulating (up-and-down) waves. When the sun shines on a shower of raindrops, the rainbow produced reveals the full spectrum (range)

ABOVE The gem sea slug or nudibranch, *Goniobranchus geminus*, from the Indian Ocean seems to glow with electric hues. This is colour for protection, as a warning to possible predators that its flesh is toxic.

of colours hidden in white light. These colours are due to different lengths of the waves, with red being the longest, and blue and violet the shortest. As the sunlight passes through the raindrops the waves are refracted (bent), with each colour bent at a slightly different angle – red is bent the least, blue and violet the most. This is how the colours are revealed, and refraction is vitally important in eyes and in nature. As individuals, we appreciate the significance and meaning of colours in so many ways, from the deepest blue sea and sky to the most beautifully coloured butterflies and flowers. However, this begs more questions. Do animals attach similar significance to colours? And without eyes to see them, do colours really exist?

Colours must be significant, since the natural world has exploited them in innumerable ways, from energy collection by the green leaves of plants, to camouflage, defence and advertising. Contrast the subdued tones of animals merging into the background with the dazzling hues of creatures as diverse as sea slugs, butterflies, coral reef fish, tree frogs and birds of paradise. Colour even lasts millions of years – the tinges of long-extinct species can be deduced from some fossils.

The human visual system has a high level of sophistication. Coupled with our capacity to investigate, learn and understand, this allows us to appreciate the central and vital roles of colour and vision in the natural world, and in our own experiences.

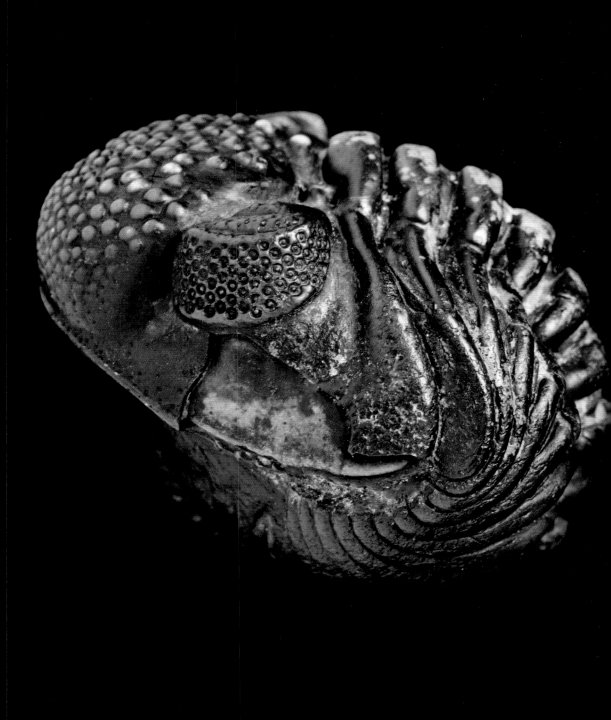

CHAPTER 1

See the light

Evolution of vision

When did eyes first appear? Before we consider this question, the term 'eye' needs some form of definition. Animals have many different light-sensing parts, and not all of them qualify as 'eyes'. An eye should be capable of receiving, responding to and comparing light rays coming from different directions to form an image, that is, to discern lines, shapes and similar features within the field of vision. This is sometimes called 'spatial resolution' or 'defining spatial detail'. The typical image-forming eye uses a lens or lenses to focus light onto a structure, such as a retina, that responds to the energy of light in some way – usually by generating nerve signals. The image-forming eye is considerably more complex than the most elementary sensors that detect either light or no light, or the next stage, comparing or qualifying the brightness of light. These simpler, more basic designs have various names and forms, such as eyespots, or photoreceptive spots or patches.

So when did eyes first appear? The immensity of prehistoric time is divided into massive spans known as eons. These are used by geologists, who study the Earth, its rocks and how they change with time, and by palaeontologists, who use fossils and other evidence to study past life and how it evolved.

First came the Hadean eon, from the time planet Earth formed some 4,540 to 4,000 million (4.54 billion to 4 billion) years ago. There is no evidence for life as we know it during this time. Next came the Archean eon, 4,000 to 2,500 million

OPPOSITE Trilobites, here *Eldredgeops rana*, were among the first major group of animals with sophisticated, multi-faceted eyes. One turret-like eye is visible in the upper-left centre.

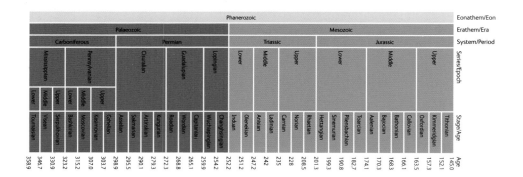

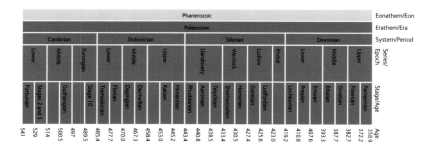

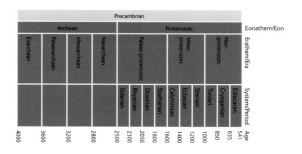

ABOVE Earth's prehistory is summarized in a series of time spans subdivided into smaller units. Eons, eras and periods refer to the spans (geochronology), while eonothems, erathems and systems refer to rock layers or strata laid down during them (chronostratigraphy).

years ago. Fossils suggest that the earliest living forms appeared during this time, in water, around 3,500 million years ago. They were microscopic single cells similar to the bacteria and cyanobacteria (sometimes known as 'blue-green algae') alive today. They would not have eyes as such, but some responded to light by capturing it as a form of energy to power their life processes. This process is known as photosynthesis.

Next followed the Proterozoic eon, from 2,500 million to 541 million years ago. Living forms evolved greater size and diversity towards the end of this span, especially animals – which by definition are multicelled rather than single-celled. There is evidence of fossils of simple animals from as long ago as 580 million years. However, their remains in the rocks are sketchy, and if they do represent animals, they would have been soft-bodied. Gradually, further evolution produced more complex creatures that probably resembled, and may have been ancestors of, groups such as the sponges, the cnidarians – which include jellyfish, hydras and corals – and the ctenophores, also known as comb jellies. Looking at living representatives of these organisms gives some clues to the early forms taken by eyes.

The most recent eon is the Phanerozoic, from 541 million years ago to the present. Fossils become more common and widespread, showing that this is when most of the creatures familiar today had their ancestral origins. Eyes made their appearance early on, and have continued to evolve since. However, eons are exceptionally long time spans. For greater convenience and accuracy, they are divided into eras, and then into periods. The Proterozoic eon consists of the Paleoproterozoic, Mesoproterozoic and Neoproterozoic eras. Most relevant for the purpose of eyes and vision is the last of these, spanning 1,000 to 541 million years ago. It is subdivided into three periods: the Tonian period, Cryogenian period and Ediacaran period.

SNOWBALL EARTH: LIFE ON HOLD

The Cryogenian period, 720 to 635 million years ago, is sometimes referred to as 'Snowball Earth'. It was a time of extreme cold over much, if not all, of Earth. Life probably struggled greatly. There are few credible fossil remains from this time. But looking at chemical remains in rocks, known as biomarkers, and working backwards using techniques such as the 'molecular clock' – which estimates how fast proteins and other molecules from living things change or evolve – suggests that the Cryogenian witnessed the appearance of certain animal groups, notably the sponges, Porifera. These very simple creatures still thrive in marine and freshwater habitats. They have no heart, muscles, guts, nerves, brain – or eyes. However, they represent a significant stage in evolution from individual, free-living cells, to groups of different cells banded together in one individual creature, carrying out different functions for the common good.

Animal beginnings

Following the Cryogenian 'Snowball Earth' period came the Ediacaran, 635 to 541 million years ago. Earth's temperatures gradually rose as glaciers retreated. Animal life began to flourish, but its members were soft-bodied and left very few fossils. This period is named after the Ediacara Hills in South Australia, where the first of these fossils were uncovered in 1946 and described for science in 1948. Similar fossils, and even whole assemblages, dating from the period have since been discovered as far apart as: Mistaken Point on the Avalon Peninsula of Newfoundland, Canada; Nama, Namibia, Africa; and the White Sea, Russia. The life forms preserved from this period are collectively known as the Ediacaran biota. Some of the fossils are difficult to interpret since they seem to represent organisms that have no clear relationships to any known groups, either living or long extinct. Others can be interpreted as possible ancestors of groups such as cnidarians – jellyfish, corals, anemones and their relatives – and the ctenophores, another group of marine invertebrates, also called comb jellies.

ABOVE Some of the Ediacaran fossils like *Charniodiscus procerus* (left) and *Fractofusus misrai* (right) at Mistaken Point, Newfoundland date to 565 million years. The site is now a Canadian Ecological Reserve and potential UNESCO World Heritage location.

ABOVE Ediacaran trace fossils are difficult to link to modern groups. Some of these burrows resemble those made by modern worms and other invertebrates, but no signs of eyes have yet been detected.

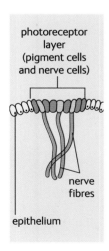

photoreceptor layer (pigment cells and nerve cells)

nerve fibres

epithelium

ABOVE A simple eye spot or ocellus can sense perhaps levels of brightness, but not direction or form an image (see page 18).

It is very unlikely that Ediacaran creatures had image-forming eyes. They may have had photoreceptors (light receivers) consisting of cells containing substances or molecules that reacted to light and triggered some sort of reaction. This would allow the organism to tell the difference between light and dark. It could be useful, for example, as a shadow detector to sense nearby movement as a passer-by cast shade over the receptor. For animals that ate life forms such as cyanobacterial mats, which gained their energy from photosynthesis, a simple eyespot could guide them towards brightly lit areas where these foods thrived. But such a device would not have been enough to form a picture of the scene and discriminate objects – to detect predators or prey, for example. Indeed, in these habitats there may have been no active predators. Life was probably slow and blind; most Ediacaran organisms were immobile. Movement would have consisted of occasionally changing position to 'mine' microbes on the seafloor or wriggling in a burrow, rather than aiming and darting as they would in the next period, the Cambrian (see below).

Basic and complex eyes

Some insight into the light-detecting parts of those ancient organisms, evolving in Ediacaran seas and freshwaters, can be gained by studying living organisms with basic, modest photoreceptors. Cnidaria, such as common jellyfish or moon jellies like *Aurelia*, have small sensing structures called rhopalia, with bundles of nerves linking them into the general net-like nerve system of the animal – there is no brain or centralized nerve 'hub' in these creatures. Rhopalia are situated in notch-like indentations on the edge or rim of the main body, the bell. Each rhopalium has

ABOVE Two of the photoreceptors called rhopalia are visible near the edge of the bell (main body) in this box jelly *Tripedalia*.

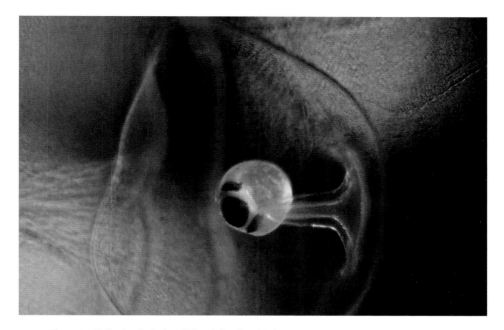

ABOVE The potentially deadly Irukandji box jelly, *Carukia barnesi*, is smaller than a fingertip. This close-up shows its camera-design eye (dark circle) with two pigment-cup eyes (brown crescents) above and below.

a patch of pigment-bearing (coloured) cells called a pigment spot ocellus, as well as gravity-sensing devices known as statoliths. Using these, the jellyfish can sense light or dark and which way up it is.

Also in the Cnidaria are the cubozoans or box jellies, named from their vaguely box-like or squarish bell. Some species are notorious for their powerful stings, with venom that can disable or even kill humans. These also have rhopalia with ocelli to detect light and statoliths to detect gravity. But in other species, two of the light detectors in each rhopalium are more sophisticated, with a camera-like design, as found in our own eyes (see Chapter 2). In these eyes there is: a covering layer of transparent cells, the cornea; a rounded lens of transparent material to focus light rays; an iris or sheet with an adjustable hole, the pupil, to control admitted light; a light-sensitive layer, the retina, consisting of photodetecting (light-sensitive) cells; a concentration of nerve fibres, probably to process signals received from the retina; and further nerve fibres linking to the body-wide nerve network. The lower, larger of these two eyes points mainly downwards while the upper, smaller one aims chiefly upwards. Using this visual equipment of more than 20 simpler and more complex eyes, box jellies respond rapidly to changing light levels and movements nearby. These are among the most energetic jellyfish, swimming purposefully with pulsations of the bell at rates of up to two metres (6.5 feet) per second as they actively pursue prey such as small fish and prawns. Presumably their visual powers allow them to identify and track potential prey.

These examples from one phylum (major group), Cnidaria, demonstrate several features concerning eyes and vision. One is that several kinds of eyes, with various grades of complexity, are found within the group and even in one animal. Also notable is the link between vision, behaviour and lifestyle. The complexity and sophistication of eye design correlates with visually guided or determined

EYES IN THE DARK

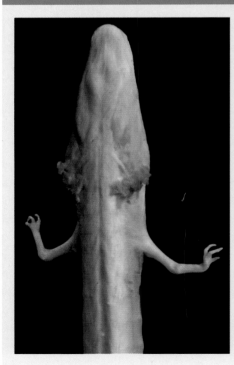

Sophisticated eyes are of little use in totally dark environments, such as caves or deep in mud. So in animals who make their homes here, and who are from groups that otherwise possess complex eyes, the eyes have gradually, through evolutionary time, regressed or even disappeared. In some instances the various stages of eye loss can be seen in a series of related creatures who live in these habitats, such as cave crickets in the insect family Gryllidae. Some species have slightly smaller eyes than typical crickets; some have more reduced versions; some have just a few photoreceptors; a few have almost no sight at all. Similar eye-reduction evolution has occurred in cave beetles and other insects, cave crustaceans such as crayfish, cave flatworms, cave fish and cave salamanders. Another always-dark habitat is the deep sea, where many fish, crustaceans, worms and others have reduced or no vision.

LEFT The degenerate eyes of the olm, *Proteus anguinus*, a cave salamander, are covered by skin yet can discriminate light levels.

behaviour, especially chasing prey or avoiding predators; this is seen in box jellies compared with other, more passive, drifting jellyfish such as *Aurelia*.

The process of evolution

In his epochal book *On the Origin of Species* (1859), English naturalist Charles Darwin, the founder of the theory of evolution by natural selection, seemed to admit that the eye posed a special difficulty: 'To suppose that the eye with all its inimitable contrivances for adjusting the focus to different distances, for admitting different amounts of light, and for the correction of spherical and chromatic aberration, could have been formed by natural selection, seems, I freely confess, absurd in the highest degree.'

However a few sentences later he suggests: 'Reason tells me, that if numerous gradations from a simple and imperfect eye to one complex and perfect can be shown to exist, each grade being useful to its possessor, as is certainly the case … then the difficulty of believing that a perfect and complex eye could be formed by natural selection, though insuperable by our imagination, should not be considered as subversive of the theory.'

Evolution is a random process driven by the mechanism of natural selection. Organisms do not make a conscious decision to alter. Variations crop up among individuals in a population due to changes in the genes (instructions in the form of deoxyribonucleic acid (DNA) for how an organism develops and how its life processes function). Any variation that is advantageous in the conditions of a certain time and place – that is, which aids survival and the ability to breed – will be favoured. Being genetic (heritable), this variation will be passed on to offspring. This is a continual, ongoing progression of adaptation. The offspring may themselves have further variations, which could be even better adapted or more beneficial and so be favoured by natural selection, and so on.

However, conditions change too. Climates warm and cool. New kinds of foods, competitors and predators evolve. Diseases come and go. These and many other shifting factors affect adaptation and survival. A trait (characteristic) that was favoured by natural selection in one time and place may no longer be advantageous elsewhere or in the future. These principles apply to features such as teeth, claws, limbs, muscle power, digestive abilities, various kinds of behaviour – and eyes.

Evolutionary stages

Studies of the connections between a box jelly's more basic visual devices – pigment spot ocelli – and its complicated camera-style eyes have led to a proposed series of steps to show how the latter could evolve from the former. At each stage the visual apparatus is a useful asset that would be favoured by natural selection, rather than being a poorly functioning 'halfway house'.

First, the basic eyespot design is a flattish layer, a spot or patch of photoreceptors – cells with pigmented (coloured) substances that absorb light, react to its energy and generate a response in the form of nerve signals. This patch is sited in the surface layer, the epithelium. Light from any direction elicits the same activity. It allows monitoring of ambient light, but in a non-directional way. Even so, this rudimentary monitoring is useful in many situations. In water, increasing dimness could be a depth gauge since it signifies sinking away from the surface; when burrowing, increasing brightness might signal coming dangerously close to the surface; and a sudden fall in ambient light might be an approaching predator.

Second, the eyespot forms a dip, cup or pit – an invagination. This adds directional information. For example, light coming in at a low angle illuminates

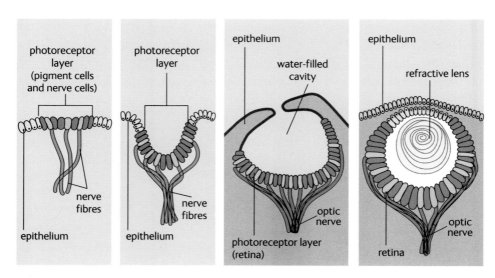

ABOVE These four stages of light receptor complexity, from simple eyespot to fully-imaging eye, all occur in molluscs.

only photoreceptors on the opposite side and near the upper rim of the cup, while light from directly above shines on most photoreceptors. The same directional information applies to shadows cast by passing objects. And the same principle applies to eyes that have a domed, bulging or mound-like design, or protrusion – an evagination. The design is useful for numerous tasks, such as basic navigation to avoid or approach objects, and identification of objects from their coarse, low-resolution size and shape.

Third, the opening to the pit or cup becomes smaller or narrower. This adds more detail to the directional information. Light from one angle stimulates only a few, and different, receptors compared with light from other angles. It also improves clarity and detail, or spatial resolution, since light from only a small area of the scene reaches only one small area of photoreceptors, instead of most photoreceptors receiving most light rays. This allows an organism to detect more detail in sizes and shapes, and to sense where a shadow that could be an enemy is positioned, so as to escape in the opposite direction.

Fourth is the cup design with the added focusing abilities of a lens. The lens bends or diffracts light rays from different distances, and at varying angles, by different amounts, which adds even more to spatial discrimination. It provides detailed, high-resolution vision of the kind employed by so many animals as they feed, breed, face competition, evade predators, avoid harsh conditions, and tackle the other trials of life.

Living examples

The four phases of eye development form a logical, plausible sequence in which one evolved from another. They have occurred on separate, independent occasions in the animal kingdom, and not always in the direction of greater intricacy. At each stage there is also 'why' – why that particular stage of eye evolution would suit the animal according to conditions of the time, and how it could furnish visually guided behaviours for natural selection to favour. And along with these increasingly complex optical abilities came steeply rising amounts of visual information from the eye, which required processing and integration to guide or direct behaviour – in other words, more nerve links and brain power.

Examples of these stages of vision occur through several main animal groups (see Chapter 2). In fact, various examples occur in just one phylum, the Mollusca. Limpets graze seashore rock and are members of the slug-and-snail subgroup, Gastropoda.

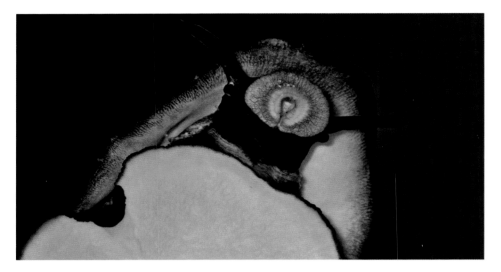

ABOVE The underside of the giant keyhole limpet, *Megathura crenulata*, reveals its mouth and two tentacles, and at the base of each tentacle, a dark eyespot.

Under the conical shell the head bears two small tentacles, each with a black eyespot at the base. These eyespots have a few pigmented cells with nerve fibres to detect light levels and have some directional capabilities between them, but little more.

A more complex stage is seen in the nautilus, a predatory oceanic mollusc in the same group as octopus, squid and cuttlefish, the Cephalopoda. The eye is a deep, rounded, cup-like chamber filled with seawater, and with a small opening called the pupil. It works in a similar way to a human invention, the pinhole camera. Such a restricted pupil means less light enters, but what does enter projects an image onto the retina – the layer of photoreceptive cells lining the chamber.

The nautilus's close relation, the octopus, has an eye very similar to that of vertebrates (backboned animals), including ourselves, although it has evolved independently. It has a camera-style design with a transparent window-like cornea, a lens for focusing, and a retina to detect detailed images. Elements of its formation and function, such as how the lens works and the layers of the retina, differ from the vertebrate eye, but the overall structures of the two designs have come to look similar – a trend known as convergent evolution. The eyes of much bigger cephalopods, the giant and colossal squid, are the largest in the animal kingdom, at more than 30 centimetres (12 inches) in diameter.

LEFT The eye of the crusty nautilus, *Allonautilus scrobiculatus*, is designed like a pinhole-camera. The small slit has no lens, it is simply an opening that allows rays through into the bowl-shaped chamber behind – where sea water can also enter.

BELOW The octopus eye, here of the common octopus *Octopus vulgaris*, is one of the most sophisticated of all invertebrates, rivalling the complexity of our own.

Several eye designs even occur in just one subgroup of molluscs, the chitons, or coat-of-mail shells, Polyplacophora. These creatures live like limpets, adhering to seashore rocks. Embedded in the shell plates are many small sensory organs known as aesthetes. In some species, these have non-pigmented photoreceptors, whereas in certain species some of the aesthetes have pigment-screened clusters of cells (eyespots), and still other species have aesthetes that bear small eyes with lenses. Chitons use information from their visual sensors to influence various kinds of behaviours, such as gripping hard onto the rock or seeking shelter. Possibly the numerous, sometimes thousands, of simple aesthetes distributed all over the shell might work as one large multipart, or compound, eye.

ABOVE Greatly magnified, the shell of the fuzzy chiton, *Acanthopleura granulata*, reveals two kinds of sensory organs: lensed eyes (dark with shiny centres) and aesthetes (pale with black centres).

A burst of evolution

Among all living animals, molluscs show the greatest variety of eye designs. Returning to prehistory, this phylum may have appeared during the Ediacaran period. But the next period, the Cambrian, from 541 to 485 million years ago, hosted probably the greatest burst of evolution the world has seen. It is often referred to as the 'Cambrian explosion'. It can be tricky to untangle how sudden and extensive the evolutionary changes really were, from the fact that fossils quickly became much more common and prominent due to the appearance of hard body parts such as shells, which preserve far more readily than soft tissues. The Cambrian marked the start of a new time span, known as the Palaeozoic era, that would last almost 300 million years.

Fossil evidence suggests that towards the end of the Ediacaran period there were only a few animal phyla (although molecular studies predict there were more than the fossils currently reveal). Yet fewer than 25 million years later – in the 'blink of an eye' on evolution's timescale – there were more than 30 phyla, and depending on exactly how phyla are defined, even approaching 40. This is similar to the number that exists today. This dramatic increase in body plans may well have been partly driven by the evolution of image-forming eyes, which started a 'visual arms race'. As an evolutionary line of predators developed better vision for hunting, so natural selection would favour any of their prey that did likewise to escape. Other advantages of vision included finding mates to breed, and communicating with others in a group.

Recent research has taken the various stages of eye evolution, and the factors that contribute to them, and calculated how long it might take to evolve from a simple photoreceptor to a complex image-forming eye, as probably happened during the Cambrian explosion. The results seem startling: fewer than 500,000 (half a million) years on pessimistic suppositions, and nearer 350,000 years on more reasonable assumptions. How so fast? Understanding the process requires study of the factors that contribute to an eye. One is its anatomy – size, shape, design and structural elements – as discussed so far and described in Chapter 2. Another factor concerns the precise materials from which an eye is constructed, that is, the light-sensitive photoreceptor molecules, the transparent proteins in the lens, and many other substances. A further factor is the blueprints to grow an eye during an animal's early development – in other words, its genes.

Genes in control

The PAX family of genes is common throughout the animal kingdom. They occur in hugely diverse groups, suggesting that they arose early in evolution and were inherited as groups separated. 'PAX' is short for 'paired box', describing the way the proteins attach to DNA. Some PAX genes carry the information to build structural substances for the body, especially the massive array of proteins that make muscle, bone, skin, nerves, eyes, noses and other parts. But the protein products coded for by some PAX genes are also controllers, or regulators, of other genes. These proteins are called transcription factors. They bind to specific sites on DNA and thereby control other genes by making them work faster, or slower, or not at all. So PAX genes can be thought of as 'genetic switches' or 'master genes'.

The PAX gene family is particularly active as an animal progresses from egg to embryo stage, when the major body parts and organs take shape. The widespread occurrence of PAX genes in different lineages of animals with eyes suggests they all share common genetic mechanisms for building their eyes. Furthermore, other functions of PAX genes suggest that this gene family appeared before eyes did, as controllers and regulators for assembling other body parts. In a sense, the PAX control system was ready to go, and already involved in other body systems, when it was co-opted for vision. With a slight change in instructions, it began to build the very earliest eyes. Then PAX genes themselves evolved to produce the variety of eyes seen today. More evidence for PAX involvement comes from mutated (changed) PAX genes that do not work properly and cause a variety of eye malformations.

THE PAX FAMILY

PAX genes are especially influential in early development. In the human, for example, the eye begins to form when the whole body is an embryo fewer than five millimetres (0.2 inches) long. There are about nine PAX genes involved in many body parts, from eyes to kidneys and the thyroid gland. In mammals, *PAX2* regulates production of a protein involved in the development of eyes, ears, the brain and spinal cord, kidney and sex organs. *PAX6* is an especially influential gene. It activates various other genes involved in the development of many eye structures, also in the nose, brain and spinal cord, and pancreas. After the eye has developed, PAX6 continues to control genes involved in its maintenance.

The discovery that the genetic capacity to build an eye is present in animals lacking eyes helps to understand how, long ago in the Cambrian, the first eyed animals evolved from eyeless ancestors – and how this happened so rapidly. The genetic part of the 'eye toolkit' was ready and waiting.

Building blocks

Genes are instructions for substances, mostly proteins, that are the structural building blocks of body parts, and that also (as explained above) control the activities of other genes. Hundreds, even thousands, of different proteins are involved in the construction of each body part. The complex eye of an octopus, fish or mammal is no exception. It incorporates many kinds of general, body-wide tissues such as muscles, nerve fibres and blood vessels, and also eye-specific tissues such as the cornea, the lens and the photoreceptive cells of the retina.

Retinal photoreceptors and similar light-sensitive cells are particularly specialized. They contain proteins called opsins which, when combined with chromophores, form visual pigments. When light energy reaches a visual pigment, the chromophore part undergoes a change in conformation (shape). This is the first part of a process known as the visual phototransduction cascade. The opsin+chromophore's altered arrangement stimulates a tiny burst of electrochemical energy that ultimately results in a nerve signal. This is conducted away along a nerve fibre, and stimulates some kind of analysis and interpretation in a nerve network or brain. The wavelength of light that each visual pigment is most responsive to is dictated by the particular amino acids that form the opsin protein. Many animals have multiple visual pigments 'tuned' to different wavelengths.

There are many kinds of opsins across the animal kingdom but only one main chromophore: retinal, or retinaldehyde, a form of vitamin A. In vertebrates the retinal cells called rods contain a partnership of opsin and retinal that forms a visual pigment known as rod rhodopsin, or visual purple. This is very sensitive to light and is responsible for vision in dim conditions; this is discussed in more detail in Chapter 2.

Opsins and chromophores are not recoverable from fossil material. However, their detailed structures can be worked out and compared in living animals. Using likely rates of evolutionary change for these types of molecules, it is then possible to delve back into their origins. As with Pax genes, it seems that opsins were already present in animals before eyes appeared. Opsins probably arose at this earlier stage to play another sensory role, such as taste or smell, and were then 'co-opted' into vision.

Eyes abound

During the Cambrian explosion, many relatives of modern marine animals first appeared in the fossil record, and they had image-forming eyes. Perhaps most fearsome was *Anomalocaris*, with some specimens well over one metre (3 feet) long – the largest creatures of the day. *Anomalocaris* has links to the vast Arthropoda 'joint leg' group of crustaceans, insects, spiders and their kin. But it is usually given its own subgroup, Dinocaridida, or 'terror shrimps'. Dinocaridids are all now extinct.

Remarkably preserved fossils of *Anomalocaris* have been studied from Emu Bay, Australia. Each of the beast's two stalked eyes was two to three centimetres (0.8 to 1.2 inches) across and had more than 15,000 individual lensed units. This multi-lensed design is known as a compound eye and broadly resembles a pincushion with many single units, ommatidia (singular ommatidium), arranged at angles so their adjacent tips form a broad dome. The design is still going strong and seen in modern arthropods such as insects (see Chapter 2). The age of these *Anomalocaris* fossils is 515 million years, demonstrating how swiftly such amazing eyes evolved. The hunter presumably used them to sight and pursue prey, indicating how far and how fast the vision-based world had come. The 'arms race' involved not only armaments, such as the spiked head appendages of *Anomalocaris*, but also the flip side of receiving sensations through the visual system.

Another smaller, less obviously vicious, Cambrian crustacean was the eight-centimetre (3-inch), shrimp-like *Waptia*. Its remains come from the Burgess Shale, a celebrated set of fossil-bearing rocks in the Rocky Mountains of British Columbia, Canada. At 510–505 million years old, the fossils of *Waptia* reveal small eyes but with the compound ommatidial construction. Even before this, at about 520 million years old, are the preservations of Chengjiang, Yunnan in southwest China. An enigmatic creature there named *Cindarella*, which may have been an early relation of the famed trilobites, had eyes that each possessed some 2,000 ommatidia.

ABOVE Recently-found fossils of the Cambrian 'super-predator'
Anomalocaris briggsi show large, complex compound eyes.

ABOVE The eyes and general anatomy of *Waptia* from Canada's
Burgess Shale rocks were remarkably similar in anatomy to
modern shrimps and prawns.

ABOVE *Hallucigenia* had spikes along its back, flexible tentacle-like legs, and simple eyes at its head end (right).

Certainly one of the oddest creatures from the Cambrian period's Burgess and Chengjiang fossil deposits is three-centimetre (1.2-inch)-long *Hallucigenia*. Its fossils show a long, flexible worm-like body, spikes, claw-ended tentacle-like extensions, and a blob or orb at one end. When reconstructed in the 1970s the spikes were placed below, as stiff leg-like appendages, and the wavy tentacles emerged upwards from its back, with the blob as the head. This was so unlike any other known organism that it received the name *Hallucigenia*, with reference to its delusional (dream-like) appearance. A reappraisal in the 1990s turned the animal upside down so that the spikes stuck up along its back and the 'tentacles' became flexible legs. However, which end was which still remained a mystery. Recent detailed research on new Burgess specimens has identified the lobe-like structure at one end not as the head but as the rear or tail, containing decayed gut fluid, or excrement. The other, narrower, spoon-shaped end is the head, now seen to have a face with a grinning ring of needle-like teeth, more teeth further inwards at the 'throat' – and two basic eyes. These were thought to detect light and dark, but not clear images. So vision of a kind was present in this worm-like Cambrian creature too. The exact classification of *Hallucigenia* has been much discussed; currently it is regarded as a relative of the living creatures known as velvet worms or onychophorans (see Chapter 2).

All-seeing trilobites

By 500 million years ago, eyes seemed almost everywhere. And almost nowhere in the fossil record are they as ubiquitous and well-preserved as in the vast group of arthropods called Trilobita, with more than 21,000 identified species through more than 250 million years of prehistory, although now all extinct. Trilobites were among the first animals to develop more complex image-forming eyes. More is known about their optics than almost any other fossil group. This is partly because the trilobite eye had a compound design with many separate

ABOVE The trilobite *Drotops*, up to 20 cm (7.9 in) in length, possessed bulbous eyes and also numerous spikes projecting from its back. Its fossils come from North Africa.

ABOVE Trilobites had a relatively conservative main body plan of three sections, left-middle-right – but an extraordinary diversity of eyes. Those of 6 cm (2½ in) long *Asaphus kowalewskii* were held up and out on stalks.

BELOW The wrap-around eyes of *Erbenochile erbeni*, with up to 500 units each, had excellent vision all around, but limited above. The 'shades' on the eye stalks perhaps protected against strong sunlight.

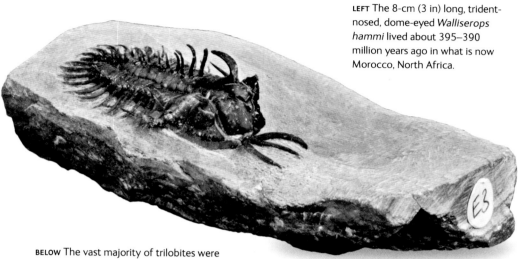

LEFT The 8-cm (3 in) long, trident-nosed, dome-eyed *Walliserops hammi* lived about 395–390 million years ago in what is now Morocco, North Africa.

BELOW The vast majority of trilobites were eyed. The lenses of some changed their refractive index (light-bending power) as light passed through, for the best possible focusing ability.

ommatidia, in which each ommatidium's lens was built from a single crystal of the mineral calcite. This mineral preserves well in the fossil record, allowing close examination of the creature's visual system.

A few examples of trilobites demonstrate their huge range of eye shapes and size. *Asaphus* lived in the Ordovician period, which followed the Cambrian, some 470 million years ago. It had cone-shaped eyes mounted on stubby stalks. *Erbenochile* appeared later, in the Devonian period around 395 million years ago. It had tall eyes wrapped around thick stalks, each with hundreds of ommatidia. It could see almost 360°, even to the rear over its back, and a shelf-like flange at the top of the eye worked like a sunshade. The extraordinary *Walliserops* lived around the same time. It had a three-pronged spear like a trident projecting from its front, and another long, curved spike sticking out from each almost spherical eye.

Colour arrives

Most of the above descriptions refer to the eye in general. One of the most useful visual features is to see colours – that is, different wavelengths of light. When did eyes evolve the ability to see these? Clues come from the types of opsin and chromophore pigment molecules already mentioned (see pages 25–26). Changes

ABOVE Reported in 2014, 300 million year old fossils of the early fish *Acanthodes bridgei* from Kansas, USA, show that rod cells and colour-discriminating cone cells were present in the eye.

in the protein structure of the opsins affect the sensitivity of visual pigments to different wavelengths of light. Each visual pigment selectively absorbs certain wavelengths (colours) of light and reflects others, which is why it seems to have a colour. This principle explains why leaves appear green: a pigment called chlorophyll in them absorbs mainly the red and blue ends of the light spectrum and bounces back the middle section – shades of green and the hues either side, pale blues and yellows. These molecular clues suggest that colour perception has very ancient origins.

In vertebrate eyes, such as our own, there are two main kinds of photoreceptor cells. Named for their shapes, they are called rods and cones. As described earlier, rod cells work well in low light levels but are not especially wavelength-selective, that is, colour-sensitive. Cone cells are of three kinds, each with pigments that absorb a narrower range of colours or wavelengths of light, being red, green and blue. Their combined information allows the discrimination in some eyes of thousands, even millions, of different colours (see Chapters 2 and 3).

When did rods and cones first appear? Detailed fossils from the Carboniferous period, 359 to 299 million years ago, suggest they were present in fish belonging to a group known as Acanthodii, or 'spiny sharks'. Members of the Acanthodii include the genus *Acanthodes* itself, with fossils widespread in North America, Europe and Australia, dating from more than 400 to about 280 million years ago. *Acanthodes* had an average length of 30 centimetres (12 inches) and sported the sharp spines in the fins that gave the group its name; its fossilized eyes indicate that rods and cones were present. Another acanthodian genus was *Cheiracanthus*, of similar size but living earlier, around 390 million years ago. The colour-discriminating pigments of fish were then carried on and evolved further by their descendants, including modern fish, and also fish who developed limbs and invaded the land, and became today's amphibians, reptiles, birds and mammals. And not only fish were evolving varied eye pigments and colour discrimination. The molluscs, and the great group of arthropods, including crustaceans in water, were developing their own colour vision systems.

With the world suddenly in focus for many species, and the ability to perceive colour, the benefits of vision became paramount. Animals had the abilities to scan for prey or other food, look out for predators and enemies and other problems, appear threatening or attractive, and communicate with others, especially of their own species. The vision-involved 'arms race' that started in the Cambrian has continued and evolved ever since.

Vertical section through tl
HORSE. *Equus cab*

CHAPTER 2

Animal eyes

Vision in different creatures

How many times have eyes evolved? As described in Chapter 1, the genetic basis and control of eye development, plus the opsins and other molecular building blocks for detecting light energy, may well have appeared just once, in the deep past and not long after the beginning of animal evolution. Using this basic 'toolkit', a wide variety of eyes has evolved at various times in different groups of animals – as the genes and molecules that produce them have also evolved. Natural selection makes each design of eye adapted (suited) to the lifestyle and needs of the creature that hosts it. Estimates for the number of times that eyes have evolved independently, in different animal groups, range from fewer than 50 to more than 100.

However, the picture is complex. As eyes and visual systems developed through time, both divergent and convergent evolution occurred. Divergent evolution is when closely related and similar organisms become different as a result of adapting to different conditions, foods, enemies and other challenges of life. This has produced a diversity of eye types that has evolved from a shared common ancestor. Convergent evolution is, in a sense, the opposite. This type of evolution leads to very similar traits in widely different, distantly related organisms, because they have adapted to similar conditions and lifestyles.

OPPOSITE A cross-section of the eye of a horse, *Equus ferus caballus*, reveals the lens and the retina lining the interior – a design common to almost all vertebrates.

So outwardly similar types of eyes have evolved independently in different groups or lineages of animals.

Convergent evolution can play tricks by suggesting a close relationship between two animal groups where there is none. For example, wings look superficially similar in birds, extinct reptiles called pterosaurs, and mammals (bats). But closer examination reveals very different internal details and evolutionarily distinct origins. The wings look the same because they are adapted to the same purpose of flight – they have converged through evolution. The same applies to vision. The eyes of vertebrates such as ourselves, and the cephalopod molluscs like the octopus, have certain similarities. But they have very different and distinct evolutionary origins.

Main designs of eyes

A major distinction among eye designs is being 'simple' or compound. In a 'simple' eye, light enters through a single opening, usually passing through a lens. So the simple eye is one visual unit, and might also be termed a 'single' eye. Several kinds of simple eyes have a camera design, named from the analogy to the human invention. The eye is built in the form of two chambers separated by a lens and light is received through a single opening. This kind of eye is found in most mammals, including ourselves.

A compound eye is 'multiple'. It receives light into more than one visual unit, that is, through many lenses, each with its own light-sensitive cells behind. The number of visual units varies according to the animal group, from just a few to many thousands, as among the insects.

The two terms 'simple' and 'compound' may be misleading, because a simple eye may have a very complex and intricate structure, as in our own eyes, whereas a compound eye may consist of only a few basic lenses, or even just a transparent covering over a few photoreceptive cells. In general, vertebrates – fish, amphibians, reptiles, birds and mammals – have simple eyes. So do some invertebrates, such as the octopus. Other groups of invertebrates, particularly the insects and crustaceans (both arthropods), have compound eyes. As mentioned in Chapter 1, the Mollusca has the distinction of possessing almost every design of eye that evolution has devised, including compound eyes.

How a 'camera' eye works

In a 'camera' eye, incoming light rays first encounter a clear, domed covering at the front of the eye, called the cornea. In fact, in our eyes the cornea has its own transparent thin covering, the conjunctiva, which is sensitive to touch and airborne substances. The protective eyelids blink every few seconds and smear lachrymal fluid (tears) to clear the conjunctival surface of dust and other items, such as vapour droplets or smoke particles. As light rays enter and then leave the cornea they are bent by the process of refraction, which happens when light passes through substances of different densities. This helps them to converge so they will focus (come together) onto the correct location inside the eye. The human cornea contributes up to 75 per cent of the eye's focusing power.

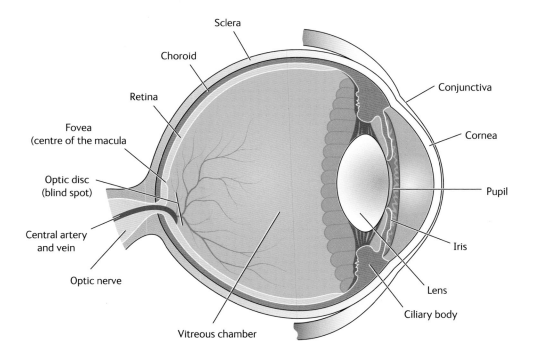

ABOVE The eye of the human, *Homo sapiens*, is typically mammalian, with a tough outer casing, the sclera, and a blood-rich choroid to nourish the inner parts.

The light rays then pass through a clear fluid called the aqueous humour, in the first of the camera eye's two chambers, the anterior segment. One of the tasks of the aqueous humour is to provide nutrients to the cornea in front of it and the lens behind. These structures must be as transparent as possible so they do not have the networks of blood vessels found in almost all other body parts. Instead they receive nutrients and remove wastes by diffusion into the aqueous humour.

Some of the light rays then pass through a hole, the pupil, in the centre of a disc of coloured muscle, the iris. The iris itself blocks the rest of the light rays: some are absorbed and others are reflected by pigments in the iris. This absorption or reflection is selective according to the wavelengths of incoming light (their colours). If an iris absorbs all wavelengths except blue, which it reflects, then the iris looks blue and the owner is said to have blue eyes. Muscles in the iris change the size of the pupil, which happens automatically by control from the brain. The pupil constricts (narrows) in bright conditions to prevent too much light entering, which could damage the eye's delicate inner parts. In dim conditions the iris makes the pupil widen, or dilate, to admit more light, which is better for seeing in darkness.

ABOVE The eye's pupil – the hole in the iris to admit light – takes various shapes; in the European adder, *Vipera berus*, it is a vertical slit.

LEFT The eye of a Himalayan brown bear, *Ursus arctos isabellinus*, shows the optic nerve leading from the rear, which carries millions of nerve signals per second to the brain.

Through the pupil, the light rays encounter the lens. This is also transparent, made of crystalline substances, and is flexible. It is convex (bulging in the middle, thinner at the edges) and surrounded by a ring of ciliary muscles. The muscles change the lens's shape, which alters the amount by which it refracts light rays. To see near objects, the muscles make the lens rounder or fatter, which bends light rays coming from the object more. For distant objects, the muscles cause the lens to become flatter or thinner, bending light rays less. These actions make sure the rays come together – converge or focus – at the correct place inside the eye, on the retina, for a clear, sharp image.

After focusing, the light passes through the second chamber or posterior segment, which forms the bulk of the eyeball's interior. This is filled with a jelly-like fluid called vitreous humour. Like aqueous humour, the jelly helps the eyeball keep its shape, provides nutrients and removes waste.

UPSIDE-DOWN WORLD?

The cornea and lens can be likened to a projector that throws a clear, detailed image of the outside view onto the retina. Because of the way light rays cross over on their way, the image on the retina is transposed (flipped) and therefore upside down and back to front. In the early stages of development, as the young brain learns to interpret what the eye sees, it soon automatically inverts the image for the 'mind's eye'.

The retina

The destination for light rays in the eye is the retina. The human retina has an area of about 1,100 square millimetres (1.7 square inches), covers 65–70 per cent of the eye's interior surface, has a thickness of 0.35 millimetres, and contains more than 100 million photoreceptor cells. It also contains: various kinds of neurons (nerve cells) that process nerve signals as they head along nerve fibres from the photoreceptor cells; a basement layer of pigmented cells to absorb stray light and prevent back-scattering or unwanted reflections; and a network of blood vessels to nourish all these parts. Curiously, the retina's nerve cells, fibres and blood vessels are 'above' the photoreceptor cells, which means they get in the way of incoming light and block some of the rays, creating a 'blind spot'. This can be regarded as an evolutionary flaw in the design of the vertebrate camera eye. (It does not happen in the cephalopod eye, where the retina is unobscured; see page 20.)

Opsins/chromophores are found in the light-sensitive cells. As described in Chapter 1 (see page 25), as they absorb light energy they change shape and trigger a series of events that send electrical signals along nerve fibres, through the layers of retinal nerve cells for preliminary processing, and then along the nerve fibres of the optic nerve to the brain, where the image is interpreted. The human optic nerve consists of about one million nerve fibres and the brain's visual centres are on its lower rear surface, called the occipital cortex.

Two eyes better

Two eyes occur in almost all vertebrates and also many invertebrates. Their placement and which portion of the surroundings they see – the field of vision – varies greatly. In primates such as ourselves and in many hunting mammals, like cats, as well as predatory creatures as diverse as owls and frogs, both eyes face largely forwards. This limits the field of view to the front but it provides what is termed binocular, or stereoscopic, vision, where the visual fields overlap, to enable accurate depth perception or three-dimensional vision – judging the distance of objects, using a number of clues. One clue is in-turning (convergence), or how much the eyeballs swivel so they both aim at an object; it is greater for nearby objects and is sensed by the eyeball-moving muscles. Another is lens shape, sensed by the ciliary muscles that change the lens to focus on near or far objects. Disparity is the difference in view between left and right eyes, which

ABOVE The Eurasian eagle owl, *Bubo bubo*, demonstrates two forward-facing eyes where visual fields overlap for a three-dimensional view (binocular or stereoscopic vision).

are compared in the brain; it is greater for nearby objects. Further clues are perspective, with convergence of parallel lines with increasing distance, and details and colours that fade with distance as blur and haze increase. In parallax, as the head changes position, near objects move more than, and across, far ones. Creatures such as monkeys, birds, snakes and lizards often move the head from side to side while gazing directly at an object, which helps to gauge distance by parallax, in-turning and disparity.

While many predators have forward-facing eyes with overlapping visual fields, many preyed-upon creatures – from antelopes to rabbits to various insects – have eyes positioned more towards the sides of the head. This gives almost

ABOVE A bottom-dweller with little need to look down, the eyes of the blue-spotted stingray's, *Neotrygon kuhlii,* provide good vision to the sides and partly above.

all-round vision to spot enemies and approaching danger. Numerous other eye positions occur. In aquatic habitats, bottom-dwelling fish and other creatures have eyes on top of the head to see to the sides and above, since looking down into the sand or mud is far from productive. Rays and stargazer fish lie on their underside and so the eyes are on top of the head. Flatfish such as plaice lie on one side of the body, and during development the eye on this side slowly moves to what becomes the upper side.

Photoreceptor cells

Most mammalian eyes have two kinds of photoreceptor cell, the rods and cones (see Chapter 1, page 33). In humans, rods are in the vast majority, numbering more than 100 million. Each rod cell is just 0.1 millimetres tall and 0.002 millimetres wide. The descriptor 'rod' refers to the upper end or outer segment of the cell, nearest the incoming light, which has a long cylindrical shape. This

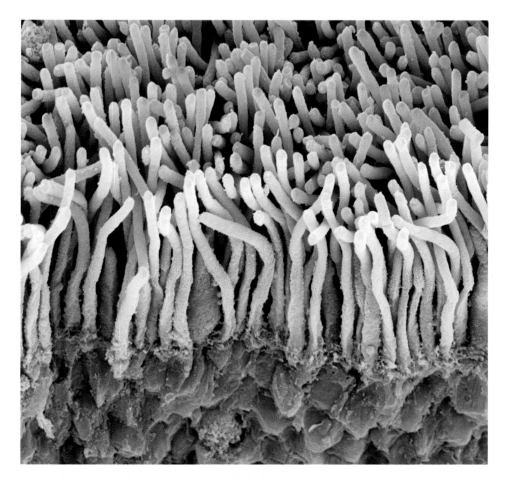

ABOVE A close-up microphotograph shows rod cells (taller, pale blue) and cone cells (shorter, mauve) packed into the retina of a human eye, *Homo sapiens*.

part is packed with a pile of structures called membrane discs, and on the sheet-like or membrane surfaces reside light-responding rhodopsin molecules, each a combination of OPN2 (opsin–2) protein and vitamin-A-derived retinal (see Chapter 1, page 25). The total number of rhodopsin molecules per rod cell is estimated at 100 million (100,000,000). Rods respond well to small amounts of light but they cannot distinguish different kinds of colours and so they 'see' in only one hue, that is, in monochrome.

Cone cells detect colours and details. In the human eye they number five or six million and are about 0.05 millimetres tall and 0.001–0.004 millimetres wide. As with rods, 'cone' refers to the upper end or outer segment of the cell, which also contains a stacked pile of membrane discs. These contain one type of cone opsin according to the type of cone, being red-sensitive OPN1LW (long wavelength), green-sensitive OPN1MW (medium wavelength), or blue-sensitive OPN1SW (short wavelength). There are fewer molecules than in a rod cell, but still many millions. Cones are concentrated in the centre of the retina, roughly opposite the lens, in an area some 5.5 millimetres across, called the macula. This is where the centre of the field of vision occurs, as the eye aims at a particular part of the surroundings. Within the macula is the fovea, only 1.5 millimetres across. Here cones are packed in tight and there are very few rods. This is the site of greatest visual acuity, detail discrimination and colour perception. However cones do not work in dim light. Humans experience this phenomenon when it is nearing darkness, and cones cease to respond; vision relies on rods alone, and the scene loses its details and colours, and 'greys out'.

As mentioned previously, cone cells in the eye allow different wavelengths or frequencies (number of waves per second) of light to be absorbed, enabling an animal to experience colour. Not all species see the same range of colours, as vision depends on the types and numbers of cones in the eye. These various chromacy conditions can be determined by analysing the opsin molecules, specifically the protein structure, in an animal's eyes, to determine the light wavelengths they absorb.

How many colours?

Monochromats are animals with one type of cone that sees in black and white, or at least, in monochrome – shades of one hue. Some mammals, including seals and sea lions, whales and dolphins (cetaceans), and certain rodents and raccoons, are monochromats. By contrast, dichromats have two types of cone, and there are many mammals in this category. They probably experience two 'dimensions' of colour, and this also occurs in certain forms of the human colour visual defect called 'colour blindness'.

Humans and our close primate relations, including other apes and some monkeys, are trichromats. So are certain marsupial (pouched) mammals such as possums and dunnarts. These have three types of cone, giving a fairly broad spectrum of colour vision. These cones are known as blue (or violet), green and red (green-yellow) cones, from the light wavelengths to which they are most sensitive. Blue cones detect short waves, perceived as purple and blue; green cones sense blue-green and green waves; reds cones respond to long waves of yellow and orange.

Tetrachromats have four types of cone, meaning that their colours are interpreted by a combination of four primary colours, rather than the three that humans possess. Tetrachromacy occurs in diverse groups, from insects to selected vertebrates such as fish, amphibians, reptiles and birds. It was also the ancestral condition inherited by early mammals, but altered genes have turned most living mammals into dichromats. A few humans have been shown to be tetrachromats, again due to changed genetic information.

Invisible light

Most of the foregoing descriptions involve what scientists term 'visible light'. This is the part of the light spectrum, or range of wavelengths, to which our own eyes respond – but it is not the whole spectrum. In fact light waves or rays are part of a much wider range known as the electromagnetic spectrum. All of these waves are composed of electrical and magnetic energy, or forces. They differ in the length of the waves and the energy they carry.

The longest waves are radio waves – an individual wave may be metres, tens of metres, or even hundreds of kilometres in length. Microwaves are shorter, from about one metre (a little over three feet) to a few millimetres long. These grade into infrared rays or waves, with lengths of around one millimetre down to 0.0007 millimetres, or going to the next smallest units, 0.7 micrometres, or 700 nanometres. (One nanometre is one-millionth of a millimetre and one-billionth of a metre.) Now comes the visible light spectrum, extending from red-colour waves 700 nanometres long, to blues and violets 400 nanometres long. Even shorter are ultraviolet waves, with lengths from 400 to 10 nanometres. Continuing the electromagnetic spectrum are the shortest waves of all, X-rays with waves of less than one nanometre, and gamma rays that are another thousand times shorter.

These electromagnetic waves are around us all the time. Radios and televisions detect the radio part of the spectrum, whereas medical X-ray machines work with waves in the X-ray range. Our eyes sense most visible light, from 390 to

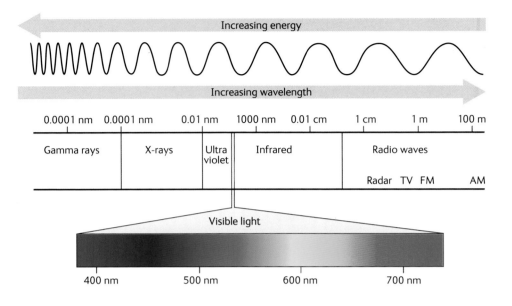

Increasing energy

Increasing wavelength

| 0.0001 nm | 0.0001 nm | | 0.01 nm | 1000 nm | 0.01 cm | | 1 cm | 1 m | 100 m |

| Gamma rays | X-rays | Ultra violet | Infrared | Radio waves |

Radar   TV   FM   AM

Visible light

400 nm   500 nm   600 nm   700 nm

ABOVE Visible light is just one small part of the spectrum of electromagnetic energy. Ultraviolet waves are slightly shorter and infrared slightly longer.

700 nanometres. Human blue cones detect waves at 420–440 nanometres, green cones 520–540 nanometres, and red cones 550–580 nanometres. However, as these lengths grade at the longer end into infrared, and at the shorter end into ultraviolet, they can be detected by the eyes of some other animals, ranging from bees and crabs to deep-sea fish.

Ultraviolet vision

The eyes of several kinds of insect, including bees and butterflies, and some birds and reptiles, can detect ultraviolet waves. Ultraviolet photography shows patterns such as lines and spots on certain flowers and fruits that are not seen in visible wavelengths of light. In some flowers the patterns point to the central area where sweet nectar is produced. Being able to detect these 'nectar guides' helps insects to find their food. In certain butterflies the wing patterns that show up in ultraviolet light are important for visual display in courtship and mating.

'SEEING' HEAT

Infrared waves carry the energy that we feel as heat. Some animals can sense this in great detail. Pit vipers are a group of snakes named from the bowl- or cup-like pit organs on the face, one either side between the eye and nasal opening. The group includes rattlesnakes, moccasins, lance heads and bushmasters. Their pit organs work in a similar way to cup-shaped eyes, but they are tuned to different wavelengths – those of infrared or heat, from around 5,000 to 30,000 nanometres (30 to 5 micrometres), which is approximately 12 to 40 times longer than visible light. The snake uses the pit organs to sense incoming heat or infrared rays, gauge their strength and direction, and work out broadly the size and shape of the object sending them – which might be a warm-blooded mouse, bird or similar meal, or a predatory snake-hunter that should be avoided. The system gives similar results to the electronic infrared 'night sights' of cameras and other night-vision equipment. The infrared (thermoreceptive) sense works along the same principles as vision but functions even in total darkness. As well as pit vipers, this sense is possessed by pythons and boa snakes, vampire bats, and insects such as some beetles and butterflies.

BELOW The pit organs of this green pit viper, *Trimeresurus albolabris*, are visible just below and towards the midline of each eye.

Mammal eyes

As described previously, all vertebrates have camera-type eyes, and different groups experience a range of acuity and colour vision. In general, these eyes are better at determining detail than movement. Eyes and vision also vary between land-dwelling and water-living species.

Primates are among the sharpest-eyed mammals, with both eyes facing forwards for stereoscopic vision. This correlates with their daylight-based, tree-dwelling lifestyles that demand fast, accurate focusing and distance judgement to move among branches, and colour discrimination to select the best kinds of fruits, leaves and other foods. It is also vital for communication with others in the group, as many primates lead social lives. Some primates are trichromats, including apes such as ourselves, Old World monkeys, and some New World monkeys. The howler monkeys of South America are trichromats. Most of their diet is made up of leaves from the rainforest canopy – choosing the most nourishing leaves from different trees and at different stages of maturation is very important, so good colour identification in the red and green parts of the spectrum is a valuable ability that natural selection would favour.

The tarsier of South East Asia is an unusual primate related to monkeys and apes. It has a head–body length of 10–15 centimetres (4–6 inches), long hind limbs, is largely nocturnal (active at night), and has evolved enormous eyes to collect as much light as possible in the conditions of dusk, night and dawn. Each eye is almost as large as the brain. This fits with a principle called Haller's rule, which states that within an animal group, bigger species have smaller brains and eyes, proportionally, than smaller species; put the other way around, as species become smaller, their eyes

LEFT Howler monkeys, *Alouatta seniculus*, are trichromats, that is, they have three different kinds of retinal cone cells for colour discrimination.

ABOVE The huge eyes of the nocturnal Philippine tarsier,
Tarsius syrichta, take up more than half of its whole head.
This primate is a dichromat, with two kinds of cone cells.

and brain become relatively larger. Tarsier eyes fit so snugly in the skull they can hardly swivel, so like an owl, the animal has to turn its head and neck to look around. The eyes are dichromatic, but rods predominate more than in other mammals, and the eye pupils can close almost to a pinpoint to keep bright daylight from damaging them.

Smaller, insect-eating bats are famous for their ability to 'see with sound', using a reflection system known as echolocation, which is akin to submarine sonar.

However, the larger bat species – fruit bats – are also largely nocturnal and crepuscular (active at twilight), but lack this ability and rely on vision. Once thought to possess only rods for low-light vision, fruit bat eyes have been found to harbour cones too. These could be useful during daytime at the communal roosting site, to spot predators and to interact socially with others in the colony.

Canines such as wolves and dogs, like most mammals, are dichromats with only two kinds of cone cell in the retina. Research shows that these distinguish yellow from blue, but not red from green – one of the forms of colour visual defect in humans. From the experiences of people affected by 'red-green colour defect', dogs were thought to distinguish blues and yellows of various hues, but what most humans perceive as reds and greens are detected by the dog eye as shades of another hue, probably a neutral such as grey. However, recent tests on pet dogs have shown that the 'shades of grey' approach is unlikely to be true and canine vision is more complex than that. This reinforces a very important point relevant to animal studies: what the eye detects cannot be transferred directly to what the creature actually perceives in its brain and conscious awareness.

The sea lion eye is adapted to focus both in air and when immersed in water, although its lens is very rounded and so is more proficient at the latter. The high numbers of rod cells function well in the dimness beneath the water surface, and the cones allow colour discrimination in the blue-green part of the spectrum. This colour ability may be an adaptation to aquatic life. Water does not treat all colours of light equally. It scatters and absorbs longer wavelengths – reds, yellows and oranges – more quickly, so these fade with depth, usually disappearing by 50 metres (165 feet) down. The short wavelengths of greens, blues and violets, which penetrate farther, perhaps to 100 metres (330 feet) are left; this is the range to which the sea lion eye is tuned. (It is also why deep, clear ocean water usually appears to be blue.) On land, the sea lion can make out clear shapes and rapid movements, but not in detail; this important point applies to many other aquatic creatures.

SHINING EYES

Pinnipeds – seals and sea lions – are among mammals that have a tapetum lucidum, helping to spot prey in the gloomy ocean. This is a shiny, reflective, mirror-like layer at the base of the retina. It reflects incoming rays back again, which gives the cone and rod cells of the retina more light to detect, but at the expense of stray rays and blurring. The tapetum lucidum is responsible for the 'eye shine' of mammals such as cats, who hunt largely at night and require good vision in very low light levels.

More vertebrate eyes

Birds in general have very large eyes for their body size, and the vast majority of species are predominantly visual. Like reptiles, birds have a clear nictitating membrane ('third eyelid'), which acts to moisten and protect the eye while allowing it to see. These membranes also occur in some mammals, such as seals, polar bears and camels, as well as amphibians and fish. Another feature unique to birds and some reptiles is the pecten. This comb- or brush-like part projects from the blood-rich layer lining the eyeball, the choroid, into the eyeball chamber. It helps with nutrition of the retina, meaning fewer retinal blood vessels to get in the way of incoming light heading for the photoreceptor cells.

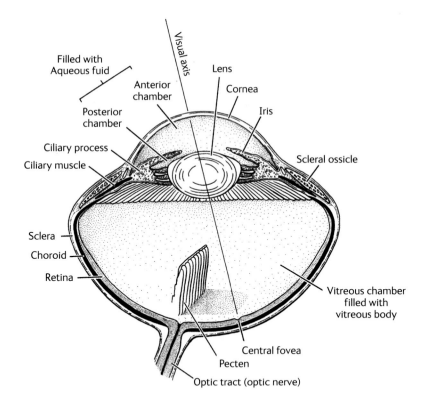

ABOVE The multi-leaved pecten is a distinctive part of the avian eye (compare with the mammalian eye, page 37).

ABOVE A Eurasian eagle-owl, *Bubo bubo*, deploys its part-transparent nictitating membrane or 'third eyelid'.

Owl eyes have taken on an unusual tubular shape with an expanded retinal end, which packs a larger area of retina into a smaller socket or orbit within the skull. The eyes of birds of prey, such as eagles, hawks and falcons, are famed for their keenness of vision. The peregrine falcon eye has two foveas linked by another region packed with cone photoreceptors, the infula. The foveas are angled in different ways so they track twisting, darting prey as the bird power-dives in its 'stoop', without the falcon having to move its head too much. Acuity tests suggest that for an object at a certain distance, the peregrine sees up to five times more detail than the human eye – or it makes out the same details of objects five times farther away.

Most reptile eyes, like those of birds, follow the standard vertebrate pattern. But again there are many specialized adaptations. For example, each of the chameleon's eyes is on a mobile cone-shaped turret that can move independently, so this lizard is able to see forwards and backwards, or up and down, at the same time. The eye has a negative lens (concave, or narrower in the middle) rather than the usual convex shape, and the cornea contributes to focusing. These features add up to rapid identification and tracking of small prey flying past.

ABOVE This panther chameleon, *Furcifer pardalis*, can simultaneously look upwards and also to the left with its independently-swivelling eyes.

ABOVE The oceanic whitetip shark, *Carcharhinus longimanus*, 'rolls' its eyes as it attacks, moving them towards the rear for protection and leaving the protective sclera visible.

Different species of fish have a great variety of eyes, all based on the vertebrate camera design but adapted to their habitats and ways of life. Shark eyes have a prominent tapetum lucidum that gives them an estimated 10 times better vision in dark conditions compared with the human eye. Many fish species have both rods and four kinds of cones (tetrachromacy) for colour vision, extending their sight beyond blues and violets to ultraviolet – these are the wavelengths of light that penetrate deepest into seawater, as explained on page 50. Deep-sea inhabitants often have huge eyes to collect any rays of light that reach their gloomy domain.

The so-called four-eyed fish has two eyes. But each eye is divided into an upper and lower portion, each with its own pupil. The upper portion is adapted to see in air, and the lower part in water, where the medium refracts light in a different way. The fish spends much of its time at the surface, watching simultaneously for danger from above and below, and for insects and similar food on the surface of the water.

'Third eye'

Some reptiles, and other vertebrates including certain amphibians and fish, have what is called a parietal eye, pineal eye or 'third eye' on the forehead or upper centre of the head. This is usually covered by a thin layer of skin and tissue, and lacks a lens or focusing mechanism to form an image, so it is perhaps better known as the parietal or pineal organ. It does have light-sensitive cells, although they are not for vision. The organ is linked to part of the brain called the pineal gland, which is involved in the rhythmic circadian (day–night) cycles of many body systems, including wake–sleep activities. The presence of light stimulates the parietal eye (named after the parietal bone region of the skull where it is located), which sends nerve messages to the pineal gland. The gland produces hormones (chemical messengers), including melatonin, which regulate physiology (bodily activities and functions), and it also has many direct nerve connections to other brain parts for the same purpose. In this way the animal's actions and behaviours keep in step with the light–dark, day–night rhythm of nature. (Its relevance to humans is described in Chapter 5; see page 112.) Lizards such as iguanas have a prominent parietal organ.

THE INSCRUTABLE TUATARA

Among all vertebrates, the parietal 'third eye' is most marked in the New Zealand reptile known as the tuatara. This creature resembles a lizard but belongs to its own reptilian group, Rhynchocephalia. The group was once varied and widespread, going back to dinosaur times, but the tuatara is the only surviving member. Early in life, as an embryo and hatchling, its parietal eye

has a cornea, lens and retina, but these structures deteriorate over a few months and become covered by skin and scales. Tuataras have many other unusual characteristics that can be regarded as 'primitive', meaning that they evolved long ago rather than recently, and in some respects these resemble the amphibian tetrapods (four-legged vertebrates) from which reptiles evolved. These features include a very basic ear, a less-evolved skull that resembles those of their fossil relatives, and a long, slow life cycle with several years between egg formation and hatching, 15–20 years to reach sexual maturity, and a lifespan that can exceed 100 years.

LEFT The site of parietal eye of the tuatara, *Sphenodon punctatus*, is just visible as a small shadowy depression between the two main eyes.

INVERTEBRATE EYES

Among the main animal groups that have evolved different kinds of eyes are the chordates, which includes the vertebrates described above – fish, amphibians, reptiles, birds and mammals. The other phyla are all invertebrates, or non-chordates. Some of these show relatively basic and conservative eye evolution, including the jellyfish and their cousins (Cnidaria; see Chapter 1, page 14), the segmented worms (Annelida) and the velvet worms (Onychophora). However, two invertebrate phyla have more sophisticated and diverse designs. As we have already seen, these are the Mollusca (chitons, snails, nautilus, octopus and their kin) and by far the biggest animal phylum, Arthropoda ('joint legs'), which encompasses crustaceans, insects, spiders and other arachnids, centipedes and millipedes, as well as the extinct trilobites and copious others. Within arthropods, insects are by far the most numerous class (subgroup), with well over one million described species. Arachnids and crustaceans are also abundant, with more than 110,000 and approaching 70,000 species, respectively. These enormous numbers of arthropods, with the vast majority having eyes (though some have lost their eyes through evolution), mean that the great bulk of animal species have vision.

Compound eye

The classic arthropod eye, as seen in insects and crustaceans, has the compound, or multiple-unit, design. As we have already seen, compound eyes consist of many individual photoreceptor units called ommatidia, each one effectively in itself a 'simple' (single) eye. It is likely that the image perceived in the creature's brain is a combination of inputs from these numerous ommatidia. We might imagine it as a 'mosaic' scene made of many tiny visual units, or even like the pixels on a screen, but how the animal itself perceives its view is difficult to determine. The main compound eye shape is based on a convex surface, so that the ommatidia point in slightly different directions and form a dome, rather like a vase packed with flowers or a fully loaded pincushion.

Each visual unit is an ommatidium, usually tall and slim, tapering to the base, and approximately hexagonal in cross section to pack closely with its neighbours (like the cells of a bee honeycomb). In bees and wasps, these ommatidia number several thousand and vary from about 0.5 to 2.5 millimetres in height or length, and 0.01 to 0.03 millimetres (10 to 30 micrometres) in diameter or width. A cockroach has around 2,000 ommatidia. Among crustaceans, the water flea, *Daphnia*, has 20–25 ommatidia per eye, the woodlouse fewer than 50. Record-holders include dragonflies and hawkmoths among insects, with 30,000-plus ommatidia, and the red rock or Sally Lightfoot crab, a crustacean, with 17,000 ommatidia.

ABOVE In this image the compound eye of the 3-mm-long
Simulian blackfly, *Simulium damnosum*, has been false-
coloured, with green for the lower or ventral portion and blue
for the larger lenses in the upper or dorsal section.

How a compound eye works

There is a specific terminology for the anatomy of the compound eye and how it works, although some terms are similar to those for other eye designs. In a typical ommatidium, arriving light first encounters the cornea, the clear covering which is continuous with the general outer body covering, the cuticle. The rays then pass through a crystal cone, or pseudocone, just beneath. The shapes and arrangements of the cornea and crystal cone vary, but in general they function as primary and secondary lenses for focusing. They are surrounded by primary or upper pigment (iris) cells that help to prevent light straying between adjacent ommatidia.

Under these structures, and forming the main length of the ommatidium, is the long, thin, transparent, crystalline rhabdom, acting as a light guide or optical fibre. This is surrounded by photoreceptor cells, known as retinula cells, analogous with the retinal cells of the camera eye. The retinula cells are long and slim like the rhabdom, usually eight in number (although ranging from six to nine or more), and lie in a ring around the length of the rhabdom, rather like a circle of rods around one central rod. Around these retinula cells is another ring of lower or secondary pigment (iris) cells, again helping to isolate the ommatidium optically.

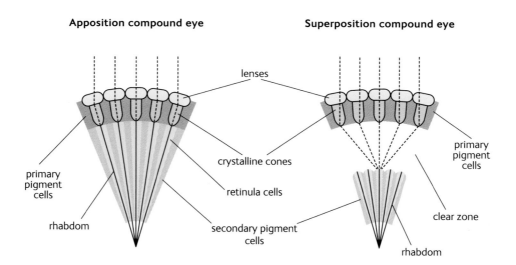

ABOVE A typical compound eye showing how the units, ommatidia, make a multi-lensed dome shape.

From the long inner surface of each retinula cell, facing the rhabdom, many miniscule finger-like microvilli extend or project towards the rhabdom at the centre of the ommatidium. The microvilli of all retinula cells are collectively known as the rhabdomere. As light rays pass along the rhabdom they stimulate light-sensitive molecules in the microvilli (as happens in the retinal cells of a camera eye) to generate nerve signals. The signals pass along nerve fibres from the retinula cells. The fibres exit the narrow, tapered base of the ommatidium and head towards the brain and main nerve system.

Apposition and superposition

Many variant designs of arthropod compound eyes, which first appeared more than 500 million years ago, have since evolved. These fall into two main groups, depending on how they form images. One is the apposition eye design, broadly as described above. Apposition eyes create one image from the many individual light sets of rays, each set entering one ommatidium. Each of these sets of light rays stimulates only, or mainly, the retinula cells in that ommatidium. The ommatidium is restricted to collecting only certain angles of light rays coming towards it, perhaps over an angle of less than one degree. This eye functions better in brighter conditions.

ABOVE Day-fliers like the large copper butterfly, *Lucaena dispar* (left) tend to have the apposition kind of compound eye. Nocturnal insects such as the Tau emperor moth, *Aglia tau* (right), mostly have the superposition design.

The other chief design is the superposition compound eye. In general the cornea–cone structures are separated by a transparent gap, called a clear zone, containing the upper parts of the rhabdoms; below are the screening lower pigment cells and retinula cells, located around only the lower parts of the rhabdoms. This allows light rays entering through one cornea–cone, or lens, to be reflected or refracted in multiple directions as it passes through the clear zone and so reach several retinula–rhabdom units below – in some cases up to 30. Apposition eyes occur in the majority of species and give a sharper, more detailed image, but they are suited mainly to brighter conditions. Superposition eyes, with their sharing or superimposing, produce a less clear view but work well in dim conditions.

EYES ON STALKS

Among the most striking arthropod eyes are those of the stalk-eyed flies, or diopsids. The eyes themselves are proportionally enormous for the animal, and each is borne on an extremely long stalk sticking out from the side of the head, which also carries the antenna. The eye span (the gap between the eyes) can be greater than the whole head–body length. One reason behind such a bizarre arrangement is probably a form of selection called sexual selection. Female stalk-eyed flies prefer to mate with males who have large eye spans, thereby having offspring with large eye spans to continue the trait.

ABOVE Male stalk-eyed flies (this species is *Achias rothschildi*) have mating displays where they face each other and display their eye span to potential female mates and rival males.

Compound eyes have many further design tweaks. Superposition can occur optically, by light, using reflection and/or refraction, or neurally, based on nerve signals, or various combinations of these. The shielding pigment or iris cells may move up and down the rhabdom according to the amount of ambient light. As they move up, light can be shared between the lower regions of adjacent rhabdoms and retinula cells, in effect turning an apposition eye into a superposition one. With many other adjustments and fine-tuning, some insects have eyes that are 1,000 times more sensitive at night than they are by day.

More arthropod eyes

Among the most complex arthropod eyes are those of mantis shrimps. In these crustaceans each eye moves independently. A single eye alone can perceive distance or depth by constantly scanning the view using three parts that are all angled to point at the same target. The middle part is a band with specialized rows of ommatidia that allow processing of polarized and ultraviolet light; an expanded number of visual pigments, up to 16, can see these features. Normally light waves undulate or oscillate at all angles – up and down, side to side, and all the positions between. In polarized light the undulations are all at the same angle – for example all vertically, or all horizontally, or all at another one angle. (Polarizing sunglasses use this feature to cut out many angles of light wave undulation and reduce glare.)

For many animals, polarized light carries useful information. For example, polarized sunlight is always at right angles to the direction of the sun, so this direction can be deduced even on a very cloudy day, for navigation by bees and other insects. The mantis shrimp senses several kinds of polarized light – even circularly polarized light, where the undulation angle rotates. The amazing vision of the mantis shrimp, apart from its use in locating food or avoiding enemies, may be in communicating with others of its kind. For instance, males display seductively to females and aggressively to other males, showing off patches of their body that shine and dazzle by reflecting polarized and other light to give many hues of colour.

In the arachnid group, spiders have 'simple' (single-unit) rather than compound eyes – the only main group of terrestrial arthropods to possess this feature. A common eight-eye arrangement is two large central or median eyes, each with two lenses and an image-forming pigment cup design, and to either side, three smaller lateral eyes with a reflective tapetum lucidum.

ABOVE The mantis shrimp's, *Odontodactylus scyllarus*, multi-hued appearance is linked to its ultra-sophisticated eyesight, which has 16 different kinds of light sensors.

A very small and distinctive group of arthropods are the limulids, or horseshoe crabs. They are not true crabs and are only distantly related to them, being more allied to spiders and scorpions. Only four species survive in their own group, Xiphosura, within the arthropods. These creatures have several kinds of eyes and other photodetectors in various positions, including the mouth and tail. The main two compound eyes each have about 1,000 ommatidia; two smaller, simpler, more central median eyes sense visible and ultraviolet light; there is also one upper central eye and two further side eyes. The compound eyes are of the apposition type that is common across the Arthropoda, and demonstrates that this was the ancestral condition for chelicerates (which include horseshoe crabs and arachnids such as spiders, scorpions, mites and ticks).

ABOVE The enormous central pair of eyes of the zebra jumping spider, *Salticus scenicus*, judge distances extremely accurately, to leap on prey or away from danger.

RIGHT Eurypterids or sea scorpions were ancient cousins of spiders and horseshoe crabs. They had a pair of compound eyes similar to those of horseshoe crabs. This is a 12 cm (5 in) long *Eurypterus remipes* from New York State, USA, with eyes clearly visible.

Mollusc eyes

In the molluscs, the nudibranchs (sea slugs) have a seemingly primitive visual set-up of small paired eyes. The eye of the opalescent sea slug *Hermissenda* consists of only five photoreceptor cells behind a lens. However there are two different types of photoreceptor; each responds to different light levels; each alters the way it responds depending on whether the actual conditions are light or dark; and each reacts differently depending on whether the other type of photoreceptor is active or inactive. So just a few light-sensitive cells can produce considerably complex information.

Scallops, which are bivalve molluscs, each have up to 100 small, basic – yet often strikingly coloured – eyes studding the fleshy mantle visible between the edge of the shell parts. Each eye has a lens, a retina, and directly behind this, a concave (bowl-shaped) reflective layer called the argentea. Light rays pass through the lens and retina, reflect off the argentea and are focused for better detection as they travel back through the retina. (The same principle with a concave mirror is used in the Newtonian (reflecting) telescope.) A single eye has poor detail

ABOVE A scallop, *Pecten* sp., part-gapes its two valves (shell halves) to reveal its two sets of small dot-like eyes.

COCK-EYED

One of the more peculiar molluscs is the cock-eyed squid group, *Histioteuthis*, found in deeper waters. The right eye is a normal size, is blue and can look down, perhaps specialized to detect the bioluminescent glow of creatures passing beneath. The left eye is much larger, tube-like, bulging, green-yellow and aims upwards to scan for food and danger.

LEFT A preserved cock-eyed squid, *Histioteuthis* sp., with the larger, protruding, up-angled left eye clearly visible.

discrimination but can detect motion, and the whole array of eyes adds up to a useful predator warning system. Other kinds of bivalves, the ark clams, belong to the family Arcidae. They, too, have many light sensors along the mantle edge, with 300 compound eyes each containing an average of 130 ommatidia but lacking lenses, and also pigment-cup designs without lenses. Image detail is relatively poor, but the wide field of view allows the clam to shut its shell rapidly when it detects an approaching object.

Worm eyes

Most annelid (segmented) worms have basic ocelli or eyespots of some kind, although a few groups have evolved image-forming eyes, and in one case compound eyes. Earthworms are annelids with tiny, basic, light-sensing eyespots, too small for us to see, spaced along the body. These detect light levels to warn the earthworm that it is approaching the soil surface, which is a less desirable place to be (unless it is breeding time).

ABOVE The feather-duster sabellid worm, *Eudistylia vancouveri,* has feeding tentacles with dozens of tiny eyes to detect danger. It then whisks its tentacles into its tube embedded in the sea floor.

Annelids include polychaete worms. Polychaete means 'many bristles' and refers to the short, stiff, hair-like chaetae along the body, giving them the common name of bristle worms. Some of the 21,700 or so species are eyeless, especially those of the deep ocean. Typical species have four pairs of unsophisticated head eyes that distinguish light from dark but probably little else. There may also be eyespots along the body. However, other species, including sabellid worms, have the most complex eyes of any annelids. Sabellids include the often beautiful fanworms, feather-duster worms and peacock worms, which are especially common in shallow warm waters. They protrude spectacular fan- or flower-like feeding head tentacles from their seabed tubes, to filter seawater for plankton food. These parts need to be exposed into the current for long periods, but they are also delicate and at risk from predators, wave-rolled pebbles, mudslides and similar threats. The tentacles are studded with dozens, in some cases more than 200, compound eyes arranged in pairs. Each eye has 40–60 ommatidia bundled into a dome shape, and each ommatidium has a lens that permits light over an angle of 10–15 degrees to enter, two pigment cells and one photoreceptor cell. The primary purpose of this elaborate set-up seems to be sensing light level rather than image formation. If a shadow is cast, even at night, the worm swiftly whisks its precious feeding tentacles down into its tube.

A separate phylum of worms contains the nematodes (roundworms) – ubiquitous creatures ranging from a fraction of a millimetre to many centimetres in length. Unlike annelid worms, the bodies of nematodes are not organized as repeating segments. Perhaps the most famous nematode in science is *Caenorhabditis elegans*, just one millimetre long, which lives almost anywhere there is rot and decay, as it consumes bacteria and other microbes. For decades *Caenorhabditis* has been a 'model organism' used for genetic and other research, in the same manner as the fruit fly *Drosophila*, and its anatomy, physiology and genetics are being uncovered in intimate detail. Although it is regarded as eyeless, recently *Caenorhabditis* was found to be able to sense light. It can detect flashes of light and wriggle away rapidly. Like the earthworm, light should be worrying to *Caenorhabditis*, because this means it is leaving the safety of the soil or feeding matter. In the research a narrow, bright beam shone onto the front end of the worm made it retreat from the light, while illuminating the tail end did the same; ultraviolet wavelengths provoked the strongest reaction. The nerve system of *Caenorhabditis* is well studied – it is known to have 302 neurons, and four kinds of these may work as basic photoreceptors, simply detecting light or dark.

A further phylum of worm-like creatures is the Platyhelminthes (flatworms). This group of more than 29,000 species includes parasitic tapeworms and flukes, and free-living polyclad species living chiefly in the sea. Some are brilliantly coloured. Many have eyespots on the head or along the edges of the body. Detailed study of these creatures has revealed intriguing features of the photoreceptor cells. In the larva (young form), one front eyespot, on the right, has three light-sensing cells that possess the finger-like extensions called microvilli, as previously described for arthropod eyes (see page 59). This is a significant feature for eye evolution, as we shall see.

The microvilli system, also called the rhabdomeric system, is common among invertebrates and is one of two basic arrangements. The other is found in vertebrate eyes, which have a ciliary photoreceptor arrangement based on

ABOVE A velvet worm, *Peripatus* sp., squirts sticky slime from under its head at a victim; the eyes that help locate the prey are at the base of each antenna.

hair-like projections known as cilia. Like microvilli, ciliary projections increase surface area, and so the numbers of light-receiving molecules, for greater visual sensitivity; however, the two systems have different origins and microstructures. This microvilli–ciliary split has evolutionary significance. Both systems occur in invertebrates and vertebrates, but for vision the invertebrates use predominantly microvilli while the vertebrates opt for cilia. Going back to the polyclad flatworm larva, its front left eye is similar to the right, but it has an additional photoreceptor cell with many cilia. As the larva grows to an adult, it loses this ciliary cell. Further studies should shed light on the evolutionary significance of these observations.

Yet another worm phylum contains the onychophorans (velvet worms), composed of some 180 species, such as those of *Peripatus*. This is a very unusual type of worm with flexible, stubby legs and are often cited as a link between legless worms of various kinds and the joint-legged arthropods. They are formidable predators, hunting down small prey such as insects on the forest floor, and disabling the victim with sticky fluid sprayed from a slime papilla near the base of each antenna. Also near the base of the antenna is a small eye. Several species of velvet worm have had the genes for their light-responding opsin molecules analysed. All possess onychopsin genes, and the velvet worm's behaviour suggests that its vision is most sensitive to blue-green light. Compared to arthropods, the details of the genes and opsins in velvet worms suggest that the two groups are closely related but separate, as parallel or 'sister' groups – in other words, one did not evolve from the other. This is another example of how eyes and vision can help to unravel evolutionary relationships.

CHAPTER 3

Making multi-hues

Colour production in nature

How many colours are there in nature? Human eyes are estimated to discriminate between more than one million shades and hues. We commonly refer to the seven prominent colours of the spectrum as red, orange, yellow, green, blue, indigo and violet. But the spectrum forms a continuous continuum (gradation), rather than being divided into discrete separate compartments each of one colour. Our experience of this vast variety has limiting factors, including our eyes, the human brain's nerve-based systems and perceptions, and human culture, memory, tradition and language, as described in Chapter 5. So our human view of 'colour' should not be taken as complete and absolute.

Natural colours are the complex result of an object's surface features, transmission properties and emission characteristics, as well as the kind of light that shines on the object and the eyes that look at it. First, an object's surface features include its physical nature and topography, whether it is rough or smooth, its hills and valleys, ridges and edges, and the dimensions of these features, which might reduce down to the scale of molecules.

Next, transmission properties include how the surface of an object allows light to pass into and though it – or not. Opaque objects do not allow light through,

OPPOSITE The stunning colours of the Madagascan sunset moth, *Chrysiridia rhipheus*, are produced in various ways, including pigment substances, and iridescence due to interference of light waves.

absorbing, scattering or reflecting it. Transparent objects such as the glass catfish allow light through almost unhindered, apart from bending or refracting the rays on the way in and out. Translucent objects such as jellyfish permit some light transmission but with scattering, so the view through is blurry and misty.

Emission characteristics involve light being produced and given off by the object itself. The ability of organisms to generate their own light and glow or shine even in otherwise complete darkness generally comes from some kind of chemical reaction, and is termed bioluminescence. Numerous kinds of animals are bioluminescent, from fireflies and glow-worms (a type of beetle larvae) on land to aquatic plankton, fish, squid and worms. Some fungi and plants also give off their own glow. Light can also be emitted in nature by phosphorescence,

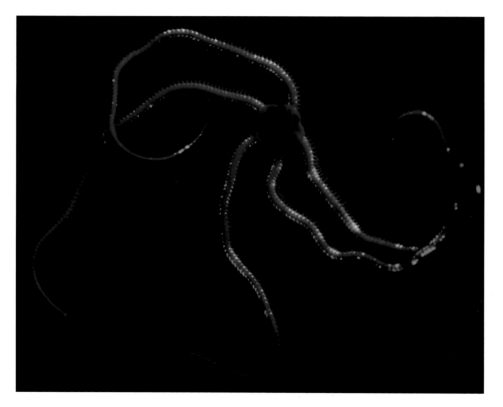

ABOVE A bioluminescent deep-sea snake brittlestar, *Ophiochiton ternispinus*, glows in the gloom with light at either end of the spectrum, red and blue.

LIGHT AND WATER

In seawater, the red end of the light spectrum is absorbed most rapidly with depth, so farther from the surface the ambient light becomes bluer. A red fish looks red near the surface because it receives all parts of the spectrum from sunlight and absorbs them all except red, which it reflects to our eyes. With increasing depth, red is filtered from the water, so there is less red available to reflect and the fish appears darker. By 30, 50 or more metres (100, 150 or more feet) deep, depending on water quality, there is so little red light that the fish has nothing to reflect and now seems dark, perhaps even almost black. But take a photograph here with a white light source, such as a spotlight or camera flash, and the fish looks red again.

which is absorbing the energy of light or other electromagnetic waves (see Chapter 2), and then sending it out again, typically as waves of a different wavelength. Biophosphorescence can emit the outward light long after the inward light energy has been received, so that an organism can 'glow in the dark' for considerable time. Biofluorescence is similar but the light is emitted over a much shorter timescale, usually only while the incoming rays with their energy are still arriving. In both these processes, biophosphorescence and biofluorescence, the emitted light has a different wavelength from the absorbed light – that is, it has a different colour.

Another aspect of colour is the ambient light reaching a surface or object. Direct sunlight carries the full spectral range, with every colour or wavelength available to be reflected, scattered, transmitted or otherwise modified. But the dimness of a forest floor receives light filtered through the green leaves of the canopy above. Sunlight reflects brightly off rocks and stones of many colours, such as red sandstone or blue gneiss, causing extra amounts of these colours to reach objects nearby.

The fundamentals of colour

In basic physics, colours are light rays or waves of different wavelengths. As discussed in Chapter 2, visible light waves have lengths of between about 390 and 700 nanometres. The shorter waves are at the violet and blue end of the spectrum, the medium-length waves green and yellow, and the longest orange and red. In terms of size, one nanometre is one-millionth of a millimetre or one-billionth of a metre. So into the diameter of this 'o' could fit 2,500 waves of violet light or 1,500 waves of red light.

These dimensions and other features of light waves are not just important in physics lessons. They are needed to understand how nature produces seemingly endless varieties of colours, shades, hues, tones, tints and tinges. These arise in two main ways, by structure and pigments. Structural colour is the production of colour by surface features – minute structured surfaces such as ridges, holes or transmitting multilayers. It is inherited by organisms, or at least largely under genetic control. Pigment colours are due to coloured substances or pigments that change the colour of light reflected from a surface by wavelength-selective absorption – soaking up some colours and reflecting others. Pigment colours can be inherited, or they can be acquired from a food source, as with pink flamingos as we will see.

Structural colour

Structural colour is the production of colour by minute structured surfaces that are fine enough to interfere with visible light. Interference involves light rays affecting each other, by changing direction and angle and combining in different ways. For example, if the waves of parallel rays of the same wavelength are completely in step, with their peaks and troughs lined up, they add together to have a stronger effect, called constructive interference. If they are totally out of step, with peaks and troughs in different places, they reduce or cancel each other out, which is called destructive interference.

In the real world, light has many wavelengths moving at all kinds of angles, reinforcing or cancelling, interfering constructively and destructively, and interacting in a multitude of other ways. For instance, blue- and red-length waves might be cancelled by destructive interference, while green waves could be emphasized by constructive interference, so the colour seen would be predominantly green. Structural colour also changes with the angle of light falling on the surface, and the angle from which it is viewed. In the above example, if the angle of the surface alters, green wavelengths could become cancelled while orange wavelengths become reinforced, so the colour would shift. This quality of colour changing with angle of illumination and angle of observation is known as iridescence.

Structural colours are produced by several mechanisms. One is diffraction. A diffraction grating is a surface structure with a repeating pattern such as periodic (regularly occurring) ridges or slits. The small size of the structures causes incoming light rays to spread out, bend and split into spectral colours, which then

ABOVE Greatly magnified scales on the wing of a swallowtail butterfly (family Papilionidae) overlap, have lobed ends and even tinier ridges – all structural elements that contribute to their colour.

interfere with each other as they travel away. A common result is a shimmering rainbow-like effect (as from a DVD or CD). Another mechanism uses a single or multiple thin-film reflector, known as a multilayer reflector. If the layers are less than one wavelength of light in thickness, they reflect different colours in different directions. With some layer thicknesses most colours leave in the same direction, producing a silver or gold effect. A further mechanism is the photonic crystal, which can be considered similar to a multilayer reflector, although often more complex, with repetitions or periodic occurrences in two or even three dimensions. Further structural colour is added by water or an equivalent, such as varnish. A very thin film of water over a surface adds another layer capable of reflecting, refracting and diffracting. This is one reason why the colours of wetted or dampened seashells 'come to life' compared with when they are dry.

ABOVE AND RIGHT Anna's hummingbird, *Calypte anna*, and the horned sungem, *Heliactin bilophus*, produce their shimmering hues with a combination of structural and pigmental colours. The exact shade changes with the angle at which they are viewed, a phenomenon known as iridescence.

These various mechanisms produce the endless iridescent, glowing hues of some animals, including the metallic sheen of certain butterflies, the gleaming bird feathers of kingfishers, peacocks, starlings and hummingbirds, and the shimmering of particular seashells, that alter as they or the observer move. The same effect is seen in plants too, including the spikemosses or lesser clubmosses *Selaginella*, certain ferns such as tropical *Microsorum*, the leaves of flowering plants such as begonias, and the glossy fruit of the African marble-berry *Pollia*. Iridescence accounts for why different people describe these colours in different ways, and why they look dissimilar in every photograph. (The same phenomenon is familiar on soap bubbles and oily patches floating on water.) Comb jellies, Ctenophora, have strips of tiny cilia arranged like teeth on a comb. These wave or beat in an undulating sequential pattern, like a sports-style 'travelling wave', and create spectacularly luminous iridescent effects.

RIGHT Clubmosses, *Selaginella* sp., seem to change from silver through pale to dark green as they are viewed in different light.

RIGHT Rainbow glimmers are produced by coordinated 'rowing' of rows of the tiny cilia along the body of a beroe comb jelly, *Beroe ovata*.

Pigment colours

Pigments are substances that cause colour by selectively absorbing, scattering or reflecting different wavelengths of light. Unlike structural colour, pigment colour does not change with the angle of incoming light and the position of the viewer. The term 'pigment' may be more familiar from industrial versions in paints and dyes, but biological pigments are found throughout the natural world. They are especially noticeable in clearly coloured material such as shells, feathers, fur and skin. A yellow pigment in canary feathers absorbs or scatters all wavelengths of white light except yellow, at about 570–590 nanometres, which it reflects. A blackbird's or crow's feather pigments absorb or scatter almost all wavelengths, resulting in the 'colour' black (arguably not a colour at all because it results from the absence of light; see Chapter 5).

There are several different classes of biological pigment. Each class produces a range of colours and the same colour can be produced by different pigments. Classes of pigments often occur in more than one taxonomic group, although a few are unique to one clade. For example betelains are found only in plants of the Carophyllales. Pigments often work in tandem with structural colour. For example, the greenfinch's feathers have a combination of a yellow pigment combining with blue structural colour to create green.

PORPHYRINS AND BILE PIGMENTS

Tetrapyrroles (bile pigments and porphyrins) are found in bacteria, plants, fungi and animals. They occur in many colours including red, green, yellow, brown and blue. For example, chlorophyll, the green pigment in plants, is a modified porphyrin. Porphyrins are also responsible for some red and brown colours found in marine seashells and red and green colours in some bird wings. Some porphyrins are implicated in the human metabolic disorders grouped under the name of porphyria, which are marked by mental and behavioural problems and skin sensitivity to sunlight.

The bile pigment bilverdin (as the 'ver' like that in 'verdant' suggests) is green and crops up in situations as diverse as the green-yellow tinge of skin bruises and the green skeleton of the garfish *Belone*. It is also incorporated into emu and other bird eggshells. Also as the 'bili' part of the name and the bruise connection suggest, biliverdin is a pigment in the green-yellow digestive fluid called bile, which is made by the liver and, if the body's metabolism (chemistry) is not working properly, carries through into urine. The usual urine pigment is a close relative to bilverdin called bilirubin.

Bilirubin and biliverdin are breakdown products of haemoglobin – a porphyrin-based substance that colours blood red. Red blood flushing through surface vessels has many visual signalling uses in nature, including our own blushes and flushes, and also the red rumps of other primates, which can be a sign of sexual approachability.

MULTI-COLOURED TURACOS

The bird family Musophagiformes (African turacos) exhibit a wide range of plumage colours, including blue, red, green and purple. The green pigment is turacoverdin, an unusual pigment among birds – most other green birds derive their colour from a yellow pigment such as a carotenoid (see page 81) coupled with structural colour in the feather itself that contributes blue, resulting in a green blend, as in the greenfinch. Another turaco pigment is turacin, providing shades of red. Both turacoverdin and turacin are porphyrins. These pigments were previously thought to occur only in turacos, but they have recently been found to occur in pheasants (Galliformes) and shorebirds (Charadriiformes).

ABOVE Turacoverdin pigment produces the green face and neck shades, while turacin gives the wingtip red, in the white-cheeked turaco, *Tauraco leucotis*.

BELOW AND RIGHT The common lobster, *Homarus gammarus* (below) has a rare blue genetic colour form or morph (right) estimated to occur in only one in a million individuals.

CAROTENOIDS

The carotenoids are among the most common biological pigments in nature, with more than 600 variants. However, only plants, certain bacteria, cyanobacteria, and a few known animal groups – including mites (cousins of spiders) and aphids (greenfly/blackfly) – are capable of making them. They vary in colour but most are yellow, orange or red because they soak up short-wavelength light. There are two chief subgroups: carotenes and xanthophylls. Carotenes are perhaps best known for making carrots bright orange. Xanthophylls can be pink, red or yellow. Along with carotenes, they lend the reds, oranges and yellows to a tree's autumn leaves; as the green light-capturing pigment chlorophyll (a chlorin pigment) is broken down and withdrawn into the tree, the carotenoids are revealed. The xanthophyll astaxanthin contributes a pink tinge to salmon flesh. Crustaceans or shellfish such as lobsters are usually mottled dull shades of browns, blues, greys and greens in life, with just a hint of red, due to a combination of various pigments including astaxanthin. Usually the astaxanthin is joined to a protein that gives a dull colour. Cooking breaks down that bond and colour changes to bright red or pink. The exceedingly rare blue form or morph of the lobster is an intense blue in life owing to astaxanthin clumping together with a protein, pulling the astaxanthin molecules closer together and causing them to absorb red light and appear blue. When cooked, the blue lobster goes pink, like any other lobster.

Among plants, tomatoes are renowned for ripening from green to red. This is a genetically programmed process in which carotenoids, especially lycopene and beta-carotene, accumulate and affect the fruit's colour, texture and flavour. Traditional selective breeding and laboratory gene modification can alter the ripening pathway and introduce another class of pigments, anthocyanins (see below), to produce tomatoes of different colours – even 'blue' varieties that are more indigo-black.

MELANINS

Perhaps the best-known biological pigments, melanins are responsible for browns or blacks in a wide range of skin, hair, fur, bird feathers, scales and squid ink. Indeed, melanins are found across most of the animal kingdom, with the possible exception of spiders. Most common are brown-black eumelanins, but there are other forms producing red, yellow, brown and tan colours. Over

ALBINISM

The condition known as albinism is usually due to a genetic change that prevents the manufacture of melanin and/or other pigments. It results in very pale or white individuals and occurs in many animal groups, including mammals, birds, reptiles, amphibians, fish, insects, crustaceans and even sea cucumbers, which are cousins of starfish in the phylum Echinodermata. In most wild habitats albinism is generally a disadvantage, as white individuals stand out prominently from the background and so are targets for predation, especially when young; the genetic causes are also linked to other health problems. Research directed at the origins and effects of albinism could help knowledge of human malignant melanoma and its possible medical treatment.

millennia, in people from environments with intense sunlight, the melanin in human skin has adapted by increasing production to darken and protect the skin. Light skin exposed to strong sunlight, especially its ultraviolet waves, begins to produce more melanin as a protective measure and so the skin darkens, giving a suntan. However skin needs a certain level of light to produce vitamin D naturally, which keeps the body healthy, which is why skin is lighter in areas of less intense sunlight, to produce more vitamin D.

Melanins are manufactured by tiny structures called melanosomes within specialized cells, melanocytes. If melanocytes begin to multiply out of control they can become the form of cancer known as malignant melanoma. The cause is usually excessive exposure to ultraviolet light, in some cases coupled with an inherited predisposition.

FLAVONOIDS AND ANTHOCYANINS

The huge flavonoid class of pigments, with more than 5,000 different kinds, occurs across a range of plants and imparts colours such as yellows, reds and reddish-blues. These substances have other roles too, from control of cell multiplication to absorbing nutrients through the roots. The anthocyanin subclass of flavonoids is responsible for certain colours in flower petals, fruits, leaves and stems – chiefly blues, purples and reds. The bright colours of flowers attract pollinators such as insects and bats, while coloured fruits attract animals that eat the fruit and disperse the seeds. Both these processes aid the plant's survival.

Acquired colours

Pigments for colour can be made by an organism itself, under instructions from its genes, or they can be acquired from elsewhere – usually, in animals, a food source. Flamingos are a well-studied example of this – not least to learn how to preserve their pretty pinks and reds when kept in captivity in zoos, wildlife parks and aviaries.

Flamingos are tall waders in the family Phoenicopteridae, related to grebes. They feed by dipping the head upside down and rapidly filtering small edible items from water through hairy comb-like structures in the bill. Different flamingo species take in a variety of cyanobacteria (blue-green algae), algae, small plants, and the larvae and adults of crustaceans such as shrimps, as well as molluscs, aquatic insects and small fish. Some of these dietary items, especially algae and crustaceans, are rich in carotenoid pigments, including canthaxanthin. The flamingo absorbs and digests these and they find their way into the feathers, legs and face, which become pink or reddish. A diet low in these carotenoids means the flamingo is paler.

ABOVE Lesser flamingos, *Phoeniconaias minor*, absorb from their food the carotenoid pigments that colour them pink.

RIGHT Gouldian finches, *Erythrura gouldiae*, of Australia occur in various natural forms or morphs, with the red face and black face morphs well represented in the wild.

Colour forms

When colour is due to pigment that is either produced or processed by an organism, a mutation (genetic change) can occur in one of the genes that produces a different colour. The presence of animals within a single species that have heritable differences in colour or pattern (or other features) is known as polymorphism. The different individuals or colour morphs may look as though they belong to entirely separate species, but they do not, and they can breed together. In some cases the colours and patterns of each morph confer their own survival value so that individuals grow up to breed and perpetuate their genes. In other instances a morph is ill adapted and may perish due to predation, disease or another threat. Alternatively it may offer protection if conditions change.

ABOVE The owl species *Strix aluco* (above left) is called the tawny owl in Britain. However, elsewhere in Europe it could be called the grey owl, from the common colour form there (right). Intermediates occur when these two morphs interbreed (above centre).

Colour polymorphism is particularly celebrated in butterflies and birds. Among parrots, wild budgerigars, *Melopsittacus undulatus*, are mostly green and yellow, with black stripes and markings, and a dark blue-green-black tail and flight feathers. However, selective breeding has become a world industry and produced budgies of almost any colour (except reds and pinks). This vast array of colour forms can be divided into two basic sets: white-based, including white, grey, sky-blue, cobalt, mauve and violet; and yellow-based, with numerous greens, grey-green, olive and yellow.

Polymorphism of colour, pattern and shape has been taken to great lengths by the great mormon butterfly, *Papilio memnon*, an Asian species in the swallowtail family, Papilionidae. As in many creatures, males and females themselves differ – this is a form of polymorphism known as sexual dimorphism. There are also differences in both sexes in the 'tails' of the hind wings, which may be present or absent, and numerous variations in colours and patterning, with more than 20 morphs in the female. To complicate matters further, some of the colour morphs are very similar to, or mimics of, other species of butterflies. The role of colour in biological mimicry is discussed in Chapter 4.

ENVIRONMENT AND COLOUR

Colours produced by biological pigments can be affected by the environment. The banded snail, *Cepaea nemoralis*, has many variations in the background colours of its shell and stripes. In general, these are paler (they contain less pigment) in hotter conditions and darker (more pigment) where it is cooler. One reason for this variation is thought to be temperature regulation, since lighter colours reflect more of the sun's light and warmth so the snail does not overheat, while dark tones absorb more to help the snail stay warmer. However, many other factors are at work, such as humidity, plant cover, and whether the snail's colour is well camouflaged to avoid predation.

The different colour and pattern forms of the banded snail represent a phenomenon known as polymorphism ('many forms').

LEFT The banded snail, *Cepaea nemoralis*, occurs in a huge variety of colour forms.

Colour change

Not all colours of an organism are constant. Chromatophores are cells that produce colours with their pigments or structural features. In some groups of animals, chromatophores can adjust their contents and structure to alter their colours. These colour changes have many purposes, including camouflage to hide from predators and other threats, warning displays to confront dangerous enemies, and communication (for example in advertising sexual status or rank in a group hierarchy).

Chromatophore-based colour change is found in invertebrates such as molluscs and crustaceans, and among vertebrates, in fish, amphibians and reptiles. Chameleons, lizards in the family Chamaeleonidae, are a well-known example, and they involve a subclass of chromatophore cells called iridophores (also found in cephalopod molluscs and some fish). In chameleon skin, iridophores produce structural colour, which combines with pigment colour. In the iridophore mechanism, the surfaces of stacks or lattices of thin crystals – nanocrystals –

ABOVE AND RIGHT Colour change in chameleons is a complex combination of structural and pigmental. Usually darker shades indicate stress or threat, and these fade to paler as the problem recedes.

LIGHT AND WATER

Cephalopod molluscs – cuttlefish, squid and octopus - are especially famed for their ability to change colour rapidly. The chromatophores can alter their size and shape, and the clumping, spreading or orientation of their pigments, to create startling changes of shade and hue that may occur at dizzying speeds, several times per second. The chromatophores are controlled by nerve signals from the animal's brain and the muscles that surround the chromatophores. As the muscles flex and contract, the animal's shape and colour patterns transform at astonishing speed. Dazzling stripes, patches and waves flash along the body, often to signal the animal's intentions during breeding and dominance struggles.

of the substance guanine interfere with light, refracting and reflecting to create the shifting colours of iridescence (see page 74). The skin has an upper layer of iridophores where spacing between the nanocrystals can be increased, which bounces back longer wavelengths of light. Under control of nerve signals and hormones, the 'default' close spacing reflects mainly blues and greens, while wider spacing yields oranges and reds. There is also a lower layer just beneath with iridophores that reflect much of the incoming light, particularly red wavelengths.

In addition, these iridophore cells contain pigments, especially yellow, which contribute to the mix and usually act with the default blues to reinforce the creature's normal greenish appearance. Like cuttlefish and other cephalopods, the 200-plus species of chameleons employ their colour transformations for a variety of reasons, from camouflage to social signalling, attracting mates, defying competitors, warning enemies, and in some cases body temperature control, as darker colours absorb more of the sun's warmth.

Resilience of colour

The resilience or lasting qualities of colours vary enormously. Pigment colour does not generally survive as well as structural colour, particularly carotenoids, which often degrade quickly and disappear soon after the organism dies. This is why the bodies of creatures such as certain fish and molluscs, so bright and dazzling in life, soon fade. However, the pigments of some organisms, including insects and flowers, may persist for months, years or decades. This is seen in museum collections of preserved insects in glass cases and flowers pressed between herbarium sheets. However, the loss of or change in colour of a specimen by techniques such as drying or spirit preservation can have an impact on the identification of the specimen. Taxonomy – the naming, describing, categorizing and classification of animals, plants and other organisms – often uses colour to identify species and groups. If this colour is lost it may become very difficult to recognize differences between species.

Some colours are resilient enough to be preserved for thousands, even millions, of years and certain pigments, such as melanins and porphyrins, may be visible in fossilized material. The 47-million-year-old fossils of the world-recognized Messel Pit, a locality near Frankfurt in Germany, include jewel beetles whose structural colours still survive with surprising vividness. This structural colour is produced by a multilayer reflector in the exocuticle, as is the case for the 'metalic' colours of living beetles. Yet even when colour is not apparent, scientists are sometimes able to identify the presence of pigments using a spectroscope, a scientific instrument that measures light and its wave properties over selected parts of the spectrum. Together with microscopic examination of preserved material, this has helped to determine colours and patterns in long-extinct animals.

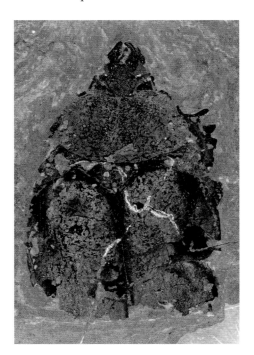 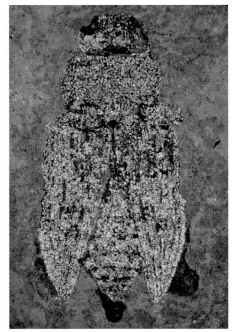

ABOVE Fossils almost 50 million years old from Messel, a UNESCO World Heritage site, still retain some of their structural colours, as in these scarab (family Scarabaeidae) and jewel (family Buprestidae) beetles.

LEFT The feathered dinosaur *Anchiornis* received its colour make-over in 2010 after studies at Yale University, USA.

Very probably the most renowned extinct animals are non-bird dinosaurs. Advanced techniques are revealing more and more of those long-gone beasts, providing amazing insights into their behaviours and lifestyles. For example, examining exceptionally well-preserved fossil skin and feathers with an electron microscope, which sees greater magnification than the traditional light microscope, sometimes identifies tiny rounded bodies, only one micrometre (0.001 millimetres) or less across. Some years ago these were thought to be the remains of bacteria or similar microbes. It has since been found that they are conserved melanosomes – the miniscule organelles (cell parts), that manufacture melanin pigments inside cells called melanocytes. An Asian non-bird dinosaur, the plant-eating *Psittacosaurus* from more than 80 million years ago, had its preserved scales scrutinized for structural colour features and for pigments, predominantly melanin. The reconstruction showed a creature covered with camouflage-like patterns of dark brown to black, lighter brown and orange-brown – entirely suitable for a dinosaur that may well have been the prey of marauding meat-eaters.

In living tissues, melanosomes come in different shapes and sizes. Those that make brown-black eumelanins, known as eumelanosomes, are shaped like sausages and are up to twice the size of the more rounded phaeomelanosomes, which make the browns and yellows of phaeomelanins. The numbers and distribution of melanosomes in fossils can therefore indicate coloration. Electron microscopes and other research tools are revealing more about the ridges, grooves, multilayers and other creators of structural colour. Every year further work extends these kinds of observations and deductions, for example, the intriguingly named time-of-flight secondary ion mass spectroscopy (ToF-SIMS), has identified preserved melanin in fossil fish eyes. This area of science is the burgeoning speciality of palaeo-coloration.

Interpreting colour from the past

The first non-bird dinosaur to receive an all-over colour scheme was *Anchiornis huxleyi*. '*Anchiornis*' means 'near bird' and this little feathered dinosaur, hardly more than 30 centimetres (12 inches) in total length, belonged to the 'raptor' (eumaniraptoran) group from which true birds arose. It lived some 160 million years ago in Liaoning Province, in what is now north-east China. Interpreting the colours of its plumage involved spotting the shapes and distribution of melanosomes in the fossils, as described above, and comparing those in the feathers of living dinosaurs (i.e. birds). Based on scientific study *Anchiornis* is now thought to have had a ridge-like central head crown of brown feathers, a black face with reddish cheek 'freckles', grey and brown over most of the body, and long arm and leg feathers with white tips studded or spangled with black, along with a grey, black-rimmed tail.

Investigating colours for the bodies and coverings of non-bird dinosaurs and other extinct species is extending to other items, such as their eggs. Studies of living birds' eggs reveal two important pigments, biliverdin and protoporphyrin (see page 78), that combine with others and the mineral calcium carbonate, 'chalk', to give different colours of eggs in different species. Close examination of fossil eggshells from dinosaurs known as oviraptosaurs detected signs of both biliverdin and protoporphyrin. The amounts and layering of the pigments suggest that their eggs were blue-green.

CHAPTER 4

Perception, deception

How animals exploit colour

Nature's uses of colours are legion. Animals exploit them to warn others that they are not an easy pushover; to hide from enemies by using camouflage; to advertise, intimidate, attract, repel, court and generally communicate, especially with others of their kind; to thermoregulate (control body temperature); to photoprotect, or shield from harmful rays; and far more besides.

Warning colours

In *The Descent of Man and Selection in Relation to Sex* (1871) Charles Darwin proposed: "With animals of all kinds, whenever colour has been modified for some special purpose, this has been, as far as we can judge, either for direct or indirect protection, or as an attraction between the sexes". It is no coincidence that many creatures that are toxic – venomous or poisonous – have bright colours. Venoms are harmful substances stabbed or jabbed into another organism using teeth, spines or stings, to damage the victim. Venomous animals use their abilities to subdue or kill prey, and to defend themselves against attackers. Poisons are substances in the body of an organism that cause harm when taken in by another creature, usually by ingesting them as food. The poisons tend to be distasteful and again they are often used in self-defence.

OPPOSITE Extremely fast and deadly venomous, the African black mamba, *Dendroaspis polylepis*, is named from its mouth's inky interior. As the snake gapes, the interior of the mouth contrasts with the pale head to warn adversaries it might strike.

LEFT The Gooty sapphire tree spider, *Poecilotheria metallica*, from one small southeast Indian forest, has a 20 cm (7¾ in) leg span and an eerie blue glow due to light-wave interference. This warns possible predators to stay clear of its irritant hairs and 2 cm (¾ in) venomous fangs.

BELOW The vivid pale blue with dark spots colouration of the blue dart-frog or arrow-frog, *Dendrobates azureus*, cautions potential predators that there is foul poison in the skin.

Numerous toxic species advertise themselves in various ways to deter predators and enemies before they themselves suffer physical harm. These warnings often include unsettling sounds such as hisses and growls, and repellent odours and scents, as well as visual signs. Many of the signals are based around contrasting patterns of bold and vivid colours, especially combinations of bright yellow, orange or red with dark hues such as brown or black. Wasps (venom), butterflies and beetles such as ladybirds (toxic flesh), poisonous nudibranchs, venomous coneshells and starfish and many kinds of fish such as lionfish, so-called 'poison-arrow' frogs, salamanders, venomous snakes, certain birds, mammals such as the foul-spraying skunk, and a host of others are involved. Humans are so vision-based that we tend to concentrate on colours and patterns, and indeed use many of them ourselves (see Chapter 5). However, ongoing research shows how, often, the other senses of sound, smell and taste are also involved in the whole warning package.

These various warning systems come under the general term of aposematism ('signal to go away'). The general idea is to tell predators and other enemies at a distance about your venom, poison or other deterrent, using dramatic warning colours, markings and other devices. An inexperienced predator may attempt an attack but soon learns that it is not 'profitable' – that is, it does not result in food, and may easily result in an unpleasant experience (pain and worse from venom, or vomiting and digestive upset after poison). The predator associates the experience with the warning signs, learns from the experience, and does not make the same mistake again. Individual victims may perish in these scenarios, but in the long run the species as a whole should benefit.

Extending this idea among several venomous, toxic or otherwise 'unprofitable' species suggests that if they all adopt similar schemes of coloration, then predators get the general idea much faster. A predator learns more quickly to avoid any creature with the common alerting signs, which benefits all of the potential prey with that coloration. When creatures copy or imitate each other's appearance, this is known as visual mimicry. If the reason for copying is to reinforce a common scheme of warning colours, it is known as Müllerian mimicry (identified in 1878–1879 by German naturalist Fritz Müller).

Among the vast assortment of warningly coloured species are the monarch butterfly and coral snake, both of North America. The monarch, *Danaus plexippus*, in the family Nymphalidae, has striking orange wings veined and edged with black, incorporating spots. Its flesh is noxious because of chemicals known as

ABOVE The 5 cm (2 in) long caterpillar of the monarch butterfly, *Danaus plexippus*, has a vivid pattern that incorporates yellow and black, a common combination of aposematic or warning colours.

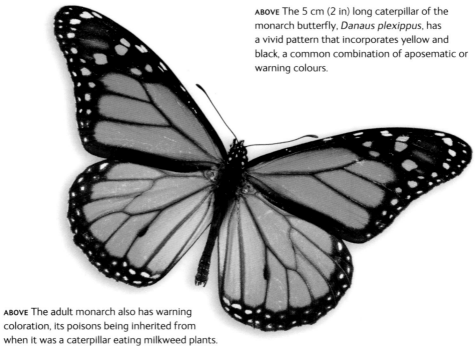

ABOVE The adult monarch also has warning coloration, its poisons being inherited from when it was a caterpillar eating milkweed plants.

cardenolide aglycones (cardiac glycosides) that have a foul taste and detrimental effects on heartbeat if consumed. The adult butterfly does not make these in its body or derive them from its food. The toxins come from milkweed plants that it ate when young – as a caterpillar. The caterpillar's flesh is likewise toxic, and it too has aposematic coloration with rings and bands of white, yellow and black.

The Eastern or common coral snake, *Micrurus fulvius*, is a member of the venomous family Elapidae, which includes cobras, adders and kraits. Its favoured prey are small reptiles and amphibians, including other snakes (even members of its own species), lizards and frogs, with the occasional fish, bird and insect. The species is marked with bright contrasting rings of red, yellow and black as a warning. The order of colours is important to humans because other snakes have similar coloration and are not venomous, and misidentification could have extremely serious consequences if a person is bitten.

It should not be forgotten, in all of these discussions, that we see warning and other colours through human eyes and with a human brain. Other creatures probably see them differently. For example, most mammals are dichromats, with only two varieties of cone cell in the retina (see Chapter 2), and so their colour perception will differ. Research has shown that, as well as the actual colours of aposematism, also important are high chromatic (colour-based) contrast and strong luminance (brightness) contrast.

THE OBLIGING MANTIS

A recent laboratory project to investigate colour and contrast used as its predator the Chinese mantis, *Tenodera sinensis*, a fiercely rapacious vision-based hunter from the mantid insect group. It has admirable compound eyes but no or very limited colour vision. Its prey were milkweed bugs, *Oncopeltus fasciatus*, in nature strikingly coloured red and black. Some bugs were reared on milkweed seeds, from which they obtained unpalatable cardiac glycosides, and some fed on sunflower seeds, which did not cause unpalatability. The bugs were painted various colours and shades to produce brightness and contrast at different levels and presented as food, and the mantis's behaviours and learning were recorded. The results indicated that the greater the brightness contrast, the more notice the mantis took, and the faster it learned whether the contrasting shades indicated tasty or distasteful prey. These are the kinds of research projects that, in their hundreds and thousands, provide our scientific knowledge about eyesight, vision and how creatures exploit colour and other features of light.

Batesian mimicry

Some creatures use a different form of mimicry than the Müllerian mimicry already described (see page 95). These are species that adopt the same warning colours as venomous, poisonous or otherwise toxic animals – but they themselves are not toxic. In human terms it might be called 'cheating'. The mimic is perfectly edible, even tasty, but its warning colours suggest that it is not. The deception is known as Batesian mimicry (identified in 1861–1862 by English naturalist Henry Walter Bates).

Both of the species described above, the coral snake and monarch butterfly, have their Batesian mimics – or so it was thought. Snakes that have similar coloration to the coral snake include certain species of non-venomous king snake (family Colubridae), such as the milk snake *Lampropeltis triangulum* (Mexican milk snake, *L. t. annulata*). At a glance the two species look similar enough to be confused. But in the milk snake the red and black bands are adjacent or touching. So when other animals learn the venomous coral snake's coloration and to avoid it, they may transfer this avoidance to the harmless milk snake too, which thereby gains protection.

Over part of its range the monarch butterfly bears an uncanny resemblance to the viceroy butterfly, *Limenitis archippus*. It was believed that the viceroy was an agreeable-tasting prey item that gained protection by imitating the unpalatable monarch. This would be a form of Batesian mimicry. However, observations and tests of birds eating the viceroy show it is likely to be unpleasant-tasting itself, over at least part of its range, depending on what its caterpillars eat. So the relationship is more likely to be Müllerian mimicry. But, as so often in nature, the overall picture is more complex. The viceroy shows visual polymorphism, or different colour forms within one species (explained in Chapter 3; see page 85). In different regions some morphs look similar to the monarch, some to the queen butterfly, *Danaus gilippus*, and some to the soldier butterfly, *Danaus eresimus*. The details of which morphs are unpalatable or not, where they occur, who mimics whom, and for what reasons, are currently being investigated.

A consequence of Batesian mimicry concerns the balance of numbers between the mimic and the creature it copies, known as the model. If mimics become too common then predators are more likely to catch models, which are tasty, and so the coloured warning effect becomes diluted or even overturned. Being polymorphic means a Batesian mimic species spreads its mimicry among several models, so reducing the dilution effect for each model species while being able to maintain high numbers itself.

ABOVE The venomous coral snake, *Micrurus fulvius*, has red, yellow and black warning colours.

ABOVE The non-venomous milk snake, this variety from Ecuador, *Lampropeltis triangulum micropholis*, bears a passing resemblance to its dangerous model, the coral snake (top).

Hiding away

Far from looking conspicuous, as in warning coloration, many creatures adopt hues, shades and patterns that blend or merge with the background, or resemble some common and unremarkable objects in the surroundings. This is the technique of camouflage, also called cryptic coloration. It provides some of the most astonishing adaptations in nature, with almost endless examples from most major animal groups. Cryptic coloration is used by predators and prey, as when a white polar bear hunts a white harp seal pup among the white snow and ice of the far north. Big cats such as tigers and jaguars merge into the patches and strips of shadow in their Asian and South American habitats. The tawny lion and the caracal, a medium-sized cat with similar coloration to the lion, both match the red-brown of the African savannas and scrubby woodlands where they live. In the sea many fish, molluscs and crustaceans such as lobsters adopt camouflage. This technique has been in use for millions of years, as evidenced by the cryptic coloration of the dinosaur *Psittacosaurus* discussed in Chapter 3 (see page 90).

Cryptic coloration often combines with shapes, behaviours, scents and other techniques, as different aspects of crypsis – the general ability to avoid observation or detection by others. The spiny turtle, *Heosemys spinosa*, of South East Asia remains still among the leaf litter of the forest floor, its brownish carapace (upper shell) matching the brown leaves around. The sharp projections around the shell's

ABOVE A jaguar, *Panthera onca*, slinks through the South American undergrowth, relying on its excellent camouflage to stay unnoticed by potential prey.

ABOVE A female adult tawny frogmouth, *Podargus strigoides*, is just discernible by her face and beak, upper right; her chick on the left has even better camouflage.

edge serve partly for protection but also match the shapes of the leaves. The tawny frogmouth, an Australian bird in the frogmouth family Podargidae, related to nightjars, hunts at night. By day it perches in the open on a tree trunk or branch. Yet its complex plumage of silvers, greys, creams, browns and blacks, with mottles and streaks, matches the bark perfectly – and the bird assumes a posture that uncannily resembles a broken-off branch stump.

ABOVE Burchell's zebras, *Equus quagga burchellii*, mingle to cause a 'jumble of stripes' that makes one individual difficult to isolate by a predator. Motion magnifies the confusion effect many times.

ABOVE Found along southern coasts of Australia, the elaborate lobes and flaps of the weedy sea dragon, *Phyllopteryx taeniolatus*, mimic the seaweeds of its inshore rocky habitat.

Disruptive coloration helps to break up or disrupt an animal's outline. The stripes of zebras fall into this category, although the full story of why zebras have stripes is more complex. Other factors may involve recognizing individuals in the herd, attracting fewer flies and other insect pests, control of body temperature, and 'dazzling' a predator as the herd gallops away so the predator finds it difficult to single out one victim.

Some creatures can alter their coloration to match different environments. Cephalopod molluscs such as certain octopus species rapidly change their coloration in a second to match their background, and the colour changes of chameleons have a camouflage aspect (both discussed in Chapter 3; see page 87). *Pleuronectes* flatfish (various species of plaice) are also colour-change artists as they adjust their spots and patches to some extent to match the seabed, although their orange spots always remain visible. But to some extent the plaice selects a background to fit its coloration, rather than changing itself to match the background.

THIS SEASON'S COLOUR

Some animals change colour over a longer timescale – with the passing seasons. A well-known example is the stoat, *Mustela erminea*, which in the north of its range moults in autumn from pale or mid-brown to white, to merge with the snowy landscape; it moults back to brown in spring. The white phase of the stoat's appearance is called the ermine. In a similar way the rock ptarmigan, *Lagopus muta*, a species in the group of game birds called Galliformes, discards its brown summer feathers for a white winter plumage. Some evidence suggests that stoat and ptarmigan moult times are happening later, or not at all in parts of their range, as a result of climate change.

LEFT This female rock ptarmigan, *Lagopus muta*, is starting her spring moult while snow still lies.

Colour and food

Humans are sometimes said to 'eat with their eyes', the implication being that food is more appreciated and even seems to be tastier if it looks more attractive. Many animals discriminate shapes and colours in foods – for example, monkeys and apes tend to choose the ripest, tastiest fruits. Some plants make their fruits particularly attractive to herbivores, who eat them and excrete the seeds to disperse them.

The attraction of colour is exploited by certain creatures to lure victims, which become their food. The alligator snapping turtle, *Macrochelys temminckii*, of southeast North America lies well camouflaged on the lake or river bottom, mouth open, and wriggles a pink, worm-shaped part of its tongue enticingly to tempt fish, crustaceans and others. These come to investigate, and the turtle snaps shut its jaws with fearsome speed and power. A similar technique is used by various anglerfish, who have a fin ray (spine) on their forehead that carries a bait or lure shaped like a flap or blob, which is bioluminescent (glowing) in some species. Again, as a curious creature comes too near, the angler's huge mouth envelops it.

Coloured feeding clues that are often invisible to humans include special markings on petals that lead to the nectar source within the flower. Some of these so-called nectar guides are in the visible part of the light spectrum, but others, like those on sunflowers, show up only in ultraviolet light – which bees and similar

RIGHT A sunflower, *Helianthus* sp., in normal daylight (left) suddenly reveals the nectar guides on its petals when viewed in ultraviolet light (right).

insects can sense. As the insect follows the guides it gathers pollen as well as nectar, then flies off to another bloom, thereby achieving pollination for the plant. In North America, the petals of Lewis' monkeyflower, *Mimulus lewisii*, change from pink, which is attractive to bees, to red, which appeals to hummingbirds. In the legume family Fabaceae (Leguminosae), coral or flame trees, *Erythrina*, have intense red blooms for a similar reason, to appeal to hummingbird pollinators.

ABOVE A 'landing spot' is visible on these flowers to insects with UV vision. This helps to ensure the right pollinator is drawn to the right flower.

Colour and sex

From molluscs, insects and crustaceans to all vertebrates – fish, amphibians, reptiles, birds, mammals – there are examples of colour used for sex. One sex of a species employs it, often along with display behaviour, sounds and scents, to attract the opposite sex for breeding. The attraction and getting-together process is known as animal courtship. In the overwhelming number of instances, the colourful one who does the attracting is the male. He is often bright and gaudy, with a dramatic performance that emphasizes his hues and other attributes. His aim is to show prospective partners that he is fit, healthy and a suitable father whose genes will benefit any offspring.

In many cases the female of the duo has a choice of courting partners, but she herself is far less spectacular. Indeed, in animal groups with parental care of eggs or babies, she is usually cryptically coloured to avoid the interest of predators and other enemies, as she feeds and guards her offspring.

Elaborately coloured and adorned males display to females in many other groups, from the blue-and-red face of a large monkey called the mandrill, *Mandrillus spinx*, to birds such as the sumptuously-tailed peacock and blue-footed booby (the blue colour due to carotenoids). Other examples are the male anole lizard's bright dewlap (chin/throat flap), the blue exhibited by the male moor frog at breeding time, the male stickleback fish's red belly display, the colourful pincers and shells of various male crabs, the wings of male butterflies that are even more dazzling than the species' females, and boundless other examples.

SONG AND DANCE

Especially extravagant courtship concerns the birds of paradise, family Paradisaeidae, of South East Asia and Australasia. In most of the 40-plus species the males are spectacularly plumaged, with bright and contrasting colours, and long trailing feathers twirled into extraordinary curls and swirls. Each male 'dances' with energetic jumps, twists and turns to show off his visual attributes to the best advantage, while uttering all manner of vocalizations such as pops, clicks, whistles and groans. These displays are among nature's most remarkable events – and in general are the result of sexual selection. A female is impressed by the most striking and competent displayer and chooses him to father her offspring. It is a form of natural selection exerted by breeding partners. The male's features are a blend of sexual selection for breeding success, and of more generalized natural selection for overall survival; the latter puts a brake on males becoming ever more alluringly magnificent, because such extravagant ornaments would make them easier targets for predators.

ABOVE A male satin bowerbird (right), *Ptilonorhynchus violaceus*, adds to the blue objects he has scattered around, and entices a female into his twiggy bower.

A twist on harnessing colour for sex is, rather than be brightly hued yourself, to transfer the attraction to colourful natural objects such as flowers, shells, stones and berries. The 20 species of bowerbirds, family Ptilonorhynchidae, of New Guinea and Australia indulge in this system. The males are relatively dull, at least compared with birds of paradise, who occupy many of the same habitats. However, the male bowerbird builds a structure called a bower or arbour from sticks and twigs, and decorates it with coloured objects to entice the female. Depending on the species, the bower varies from a relatively untidy pile of twigs and leaves, to a carefully constructed and shaped edifice in the design of a cone, hut or tent, or as an avenue, walkway or similar. The male gathers and painstakingly positions sometimes hundreds of objects selected for size and colour, such as leaves, petals, shells, feathers, pebbles, fruits and berries, and he also dances to charm the female. In modern times he may even incorporate suitable fragments of human trash such as bits of plastic, glass, clothes pegs, pens and even shotgun and rifle shells. The satin bowerbird, *Ptilonorhynchus violaceus*, specialises in blue objects and has been known to favour blue ballpoint pen tops, blue bottle tops and blue drinking straws. As in many other species, most female bowerbirds are inconspicuous in coloration, in speckled greens and greys, as they go about nest-building and chick-rearing with no input from the male.

Colour and temperature

Physical sciences show that, compared with lighter items, darker objects absorb more of the sun's radiation – light, heat and other waves – and become warmer. Various kinds of animals use colour to thermoregulate. Dark versus light colours are part of the inherited variation in the banded snail described in Chapter 3. On the Galapagos Islands the only truly sea-frequenting lizard, the marine iguana, *Amblyrhynchus cristatus*, can darken its skin to absorb more of the sun's heat while on land, before it dives perhaps as far down as 25 metres (a little over 80 feet) into the cold water to feed on seaweeds (marine algae). On its return the chilled iguana is slow and vulnerable until it has sunned again and restored its body temperature. Some turtles adopt a similar system as their limbs darken and stretch out to absorb warmth, then go paler and retract to prevent overheating.

Colour, protection and strength

In the mammalian species *Homo sapiens* (we humans), light-skinned individuals usually show a longer-term darkening over days and weeks, to shield the skin from the sun's damaging rays (see page 82). This is an example of colour being

ABOVE Chilled after diving, a marine iguana, *Amblyrhynchus cristatus*, has darkened skin to soak up more of the Sun's warmth.

used for photoprotection. A similar protection utilizing carotenoid pigments is found in flies such as houseflies and mosquitoes, and in aquatic invertebrates such as *Daphnia* and its crustacean cousins, the copepods.

Animal pigments have uses other than for their colour. They are involved in structure and strengthening. For example, in insects, melanin pigments in the body's outermost layer, the cuticle, help to strengthen, stiffen, reduce abrasion, repel germs and fortify wound healing.

Green world

Plants use colour in the sense that their photosynthetic pigments absorb light energy to power their lives. By far the most common of these is chlorophyll. Algae or seaweeds are different colours so they can absorb light at different depths of water – usually green algae higher on the shore, then browns (which harbour the pigment fucoxanthin) towards low tide, grading to red algae in deeper waters. Here the red algal phycobiliprotein pigments such as phycoerythrins are efficient at absorbing blue light, which is the only wavelength to penetrate to such depths (see page 50).

Virtually all animal life depends directly or indirectly on plant pigments to harness sunlight energy and turn it into energy-rich living tissues. The plant tissues are eaten directly by herbivores, and energy passes indirectly through them to carnivores.

ABOVE Very low tides reveal *Laminaria* seaweeds; their brown colour is adapted to absorbing light at depth.

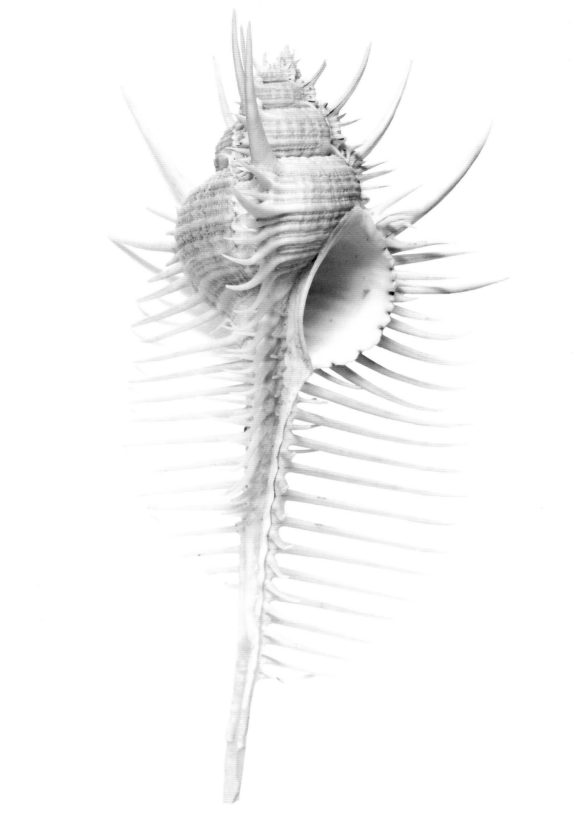

CHAPTER 5

Our rainbow world

Human experience and exploitation of colour

Humans are a relatively plain species. Our skin colour has adapted to suit particular environments, but we do not use our natural skin and hair colours in the ways that many animals do – for example, for camouflage, or for warning of venom or poison. However, in today's world, people exploit 'added colour' in an immense variety of ways. For the body we choose coloured clothing to wear, coloured cosmetics, dyed hair, ornamented nails, skin paints and tattoos. Fashions of colour come and go. We select colours and patterns to decorate all manner of locations including our homes, community areas, public spaces and buildings, offices and other workplaces. Coloured commercial products range from smartphones and kitchen gadgets to cars and aircraft. Colours are associated with the slickest brand names, global trademarks and corporate identities. Colours of foods and drinks, natural and artificial, are used by their producers and advertisers in all kinds of calculating ways to stimulate, if not taste buds, then an urge to buy.

Humans of various cultures have also taken a leaf out of nature's book with the common use of the colours red, yellow and green. Red tends to signify 'stop' or 'danger', yellow applies to warnings and cautions (both common in warning coloration, as we have seen), while green signifies 'go' or 'satisfactory'. These are familiar in all manner of everyday situations, from traffic lights, hazard signs in industry and coloured icons on electronic screens, to bright levers and knobs on children's toys – as learning the significance of colour starts early.

OPPOSITE For centuries, murex sea-snails such as the Venus comb, *Murex pecten*, have been collected to obtain the valuable dye Tyrian purple.

LIGHT AND BODY RHYTHMS

For as long as humans have existed we have interacted with natural colour and light – and not only by vision. For example, the body has a natural circadian – 24-hour – cycle of activity and biorhythms, synchronized or entrained to the day and night regularity of light levels. The retina of the eye sends some of its nerve signals to a small part at the lower front of the brain called the suprachiasmatic nucleus. This is just above the main nerves dealing with vision and is the 'body clock'. Rising and falling light levels affect its activities through nerve and hormone mechanisms, including the pineal gland that produces the 'sleep hormone' melatonin (see Chapter 3; page 55). Alertness, wake and sleep, body temperature, digestion, urine production and many other bodily functions are involved in these rhythms, as people know who travel rapidly through time zones and experience jet lag, or who switch from day work to night shifts.

Colour has a strong cultural significance that does not exist elsewhere in the animal kingdom, again with a huge range of applications: national flags and traditional dress, football and other sports clubs, special-interest societies and community organizations.

Interactions with colours affect our moods, emotions, motivations and expectations, and are manipulated in various subtle ways by designers, artists, stylists and advertisers. Thus, in some cultures, the reds, resembling blood, are regarded as physical, arousing, strong, powerful, even aggressive; 'hot' oranges, as in flames, may elicit strong emotional responses including enthusiasm and encouragement; yellows, being reminiscent of sunshine, can impart happiness and energy; greens signify nature with harmony and balance, as well as safety; blues, as in the sky and sea, are calmer, more tranquil and more intellectual; and purples and violets are associated with luxury, power, wealth and extravagance (see below).

Colours from nature

Since the time before recorded history, animal and plant pigments have been collected and exploited by people for body ornamentation, fabric dyeing, painting and other artistic activities, and applied to utensils, tools and weapons. Recent discoveries suggest that as long as 50,000 years ago our close cousins the Neanderthals, *Homo neanderthalensis*, manufactured pigmented substances that were perhaps used as 'body paint' or to colour body adornments. Seashells from this time found in Murcia, southern Spain, have preserved pigmented residues from mixing or storing the paint, including yellow, red and shiny black. Their date is well before signs of modern humans were present in the area – but a time when Neanderthals were known to have been there.

ABOVE King scallop shells, *Pecten maximus*, from the Spanish Cueva de los Aviones site show signs of various pigments concocted and stored by Neanderthal people.

Natural colours took on cultural and ritual significance. Tyrian purple is well known from the civilizations of Ancient Phoenicia, Greece and Rome. With strong anti-fade properties, it goes by several other names, such as imperial purple and royal purple, signifying its status. The pigment is collected from predatory sea-snail molluscs of the murex group, Muricidae, specifically from the body parts known as hypobranchial glands, which produce slimy mucus and other fluids. Unusually for animals, the pigment contains the chemical element bromine, and the snail uses it to subdue prey – mainly other shellfish. In the past, periodic scarcities of these sea snails only served to heighten the dye's precious rarity and encouraged more collecting from the wild.

ABOVE Cactus-feeding cochineal scale insects, *Dactylopius coccus*, are the source of a strong red dye and are also used for the homeopathic remedy Coccus Cacti.

Cochineal 'beetles' are scale insects, in the superfamily Coccoidea, of the true bugs Hemiptera, that have also been used through history as the source of a pigment and dye. This is a deep intense red, termed cochineal, carmine or natural crimson. It has many and varied uses, from paints, inks and fabric dyes to cosmetics and food colourings. The pigment is based on a chemical called carminic acid, obtained by powdering the dried insects and mixing them with aluminium-containing chemicals. Scientific study shows that the pigment is very efficient at reflecting only wavelengths of light longer than 605 nanometres – the red end of the visible spectrum.

ABOVE A big-fin reef squid, *Sepioteuthis lessoniana*, ejects ink to conceal its escape. The sepia ink's melanin pigments have been used by people for millennia.

Further examples include sepia, a brownish-red pigment derived from the 'ink' made by various cephalopod molluscs, especially squid and cuttlefish, and originally derived from a genus of the latter with the scientific name *Sepia*. The ink is made in an ink sac and is a very concentrated form of melanin. When threatened, the creature squeezes it from the sac into the surrounding water as a dark cloud, behind which the mollusc makes its getaway. Natural sepia has been used for paints and dyes since ancient times, and it gave its name to photographic techniques that make an image browner or some similar tone, which seems aged and 'warmer'.

NAMES OF COLOURS

Traditional Western perception and language names the 'seven colours of the rainbow': red, orange, yellow, green, blue, indigo and violet (see Chapter 3, page 70). But in reality, the spectrum is a continuum of hues. Different human cultures perceive colours differently, with more language terms and greater discrimination for those that are important in the local environment. The Himba people of Namibia have a colour 'serandu', which is what many English-speakers would name as a range of reds, pinks and oranges. The Himba term 'zoozu' encompasses varied dark hues that in English would be labelled dark blues, dark greens, dark browns, dark reds, and even black. The Shona people of Zimbabwe identify four main colours in the spectrum: 'cipsuka' is red-orange or blue-purple, 'cicena' is yellow to yellow-green and 'citema' is greenish-blue. Australian aboriginal languages have more, and more detailed terms, for reds, browns and yellows than English speakers would use – terms coming from the predominant colours of their arid environments.

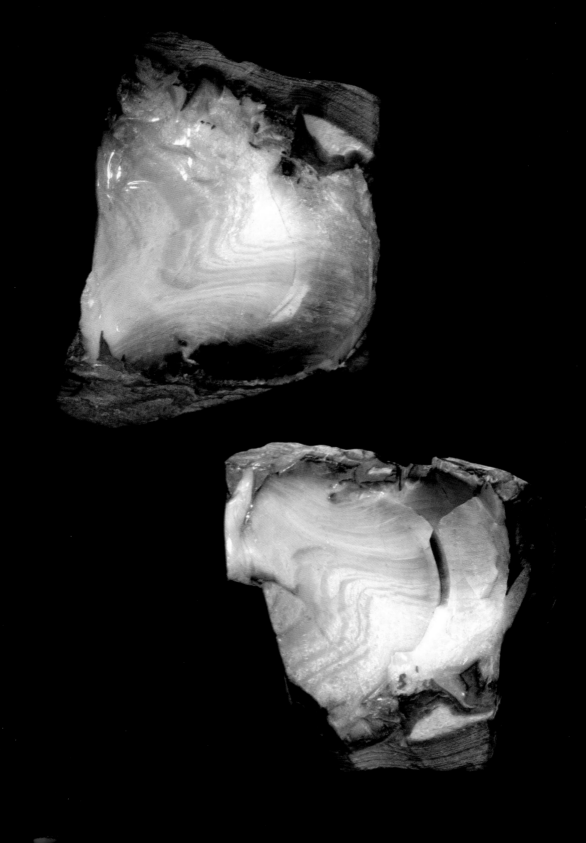

Nature first

In modern technology, many uses and exploitation of colour have been inspired by the natural world – although sometimes nature got there first, and we found out only later. Artificial versions of the opal gemstone are notoriously difficult to manufacture. However, it has been discovered that the body casing of the weevil *Pachyrhynchus argus* (a kind of beetle) from northeast Australia has a similar micro-optical structure to opal, with minuscule spheres creating metallic, glossy and milky shades visible from any viewing direction. The system acts as a type of photonic crystal, as explained previously (see Chapter 3, page 75). It could lead to improved synthetic opals as well as fibre-optic applications in communications and electronics.

Manufactured lenses are at risk of spherical aberrations, where incoming light rays near to the periphery or edges are bent, or refracted, at slightly different angles compared with those in the lens's centre. The result is that not all of the rays passing through all areas of the lens come to the same point or focus on the other side. Remedies include using a combination of convex and concave lenses, known as a compound lens, and adjusting the material from which the lens is made so it grades in bending or refractive power from centre to periphery. However, a subgroup of trilobites including the genus *Phacops* were doing this hundreds of millions of years ago, during the Ordovician to Devonian periods (485 to 359 million years ago). In each visual unit or ommatidium of the eye was a relatively spherical lens, and near it a bow-shaped lens with a wavy surface. Together the two lenses, or 'doublet' – preserved since they were made from the mineral calcite (see Chapter 1, page 32) – corrected for spherical aberration.

The sea mouse, *Aphrodita*, is a kind of predatory marine worm in the polychaete group, related to the fanworms mentioned in Chapter 2 (see page 67). Usually 15–20 centimetres (6–8 inches) long, it lives as deep as 3,000 metres (around 2 miles) down and has long, hair-like, fibrous projections on its upper surface. These work by structural coloration, as hollow nanofibres in hexagonal arrays. They are similar to multi-dimensional photonic crystals produced by industry since the 1980s for applications in optics and electronics, including advanced optical fibres (fibre-optics) for communications. The sea mouse's hair colour

OPPOSITE Natural opals produce their delicate shades by repeating 'nanostructures' at the scale of single atoms and molecules, termed photonic crystals.

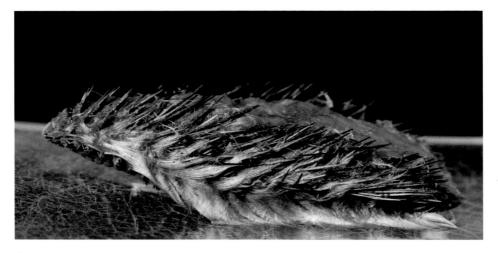

ABOVE The sea mouse's striking colours, created by its nanotube 'hairs', gave it the scientific name of *Aphrodita* from the Greek goddess of beauty and love.

varies from deep red to pale blue-green, with colour functioning probably to warn off predators. In another example, the 'mirror-box' optical systems in the compound eyes of prawns, shrimps and lobsters, have also recently been studied with a view to improving wide-angle lenses so they can deal with wavelengths ranging from very short X-rays to long infrared, rather than only one narrow portion of wavelengths.

White than white?

One of the world's whitest natural objects is a beetle, a genus of the scarab family known called *Cyphochilus*. It is whiter than any paper and almost any other human-made material, due to structural rather than pigment colour (see Chapter 3, page 78). Research shows that the scales on the beetle's covering reflect all colours of light equally in the most random way, so that any angle of view receives all of the wavelengths, which are perceived as white. Another startling fact is the thinness of the long, reflective, overlapping surface scales – just five micrometres (0.005 millimetres or five-millionths of a metre) in thickness. The effect is based on nanofilaments and microfibres of the substance chitin, a common structural material in the animal world. The filaments are randomly orientated but of carefully balanced sizes and spacing. Similar incredibly thin, light microstructures could be used in industry to make paper, screens, paints and many other products 'whiter than white' – even toothpastes.

Computer modelling of the beetle's system suggests that incorporating gold and other atoms in specific ways could do the opposite and make the blackest of all materials. One version would absorb not only visible light waves but also ultraviolet, and convert all these to infrared or heat, leading to possibilities such as more efficient solar panels. Tinkering with the nanostructure might even produce a variety that could take in all visible light waves but give out just one pure colour, that is, a single wavelength.

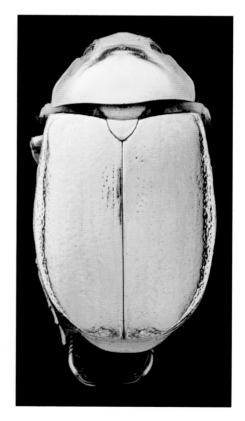

ABOVE The production of metallic sheens by jewel scarab beetles, *Chrysina* sp., is being investigated for possible use in optical devices.

HEARING COLOURS

Neil Harbisson (1982–) was born without colour vision so that his eyes see a naturally monochromatic or greyscale world, rather like old-fashioned 'black-and-white' movies. Growing up in Spain, he studied arts and music, and became involved in nature conservation. Partly as an experimental arts-based project, in 2004 Harbisson underwent surgery to implant devices in the back of his skull with an antenna-like extension over his head. The devices transmit into his head various vibrations that he hears with his inner ears, as sounds of various frequency or pitch. The antenna's sensor can be adapted for vision and colour, including infrared and ultraviolet, and for radio frequencies such as Bluetooth and telephone communications. In this way Harbisson associates sound frequencies with light wavelengths and so 'hears colours' in great detail. Active in the arts and media, and a regular at campaigns and festivals, Harbisson has brought a new dimension to the meaning and significance of colour.

Vision and other senses

Technology has allowed us to see what our own eyes cannot sense, with devices that detect infrared, ultraviolet and other wavelengths, and then convert the images to visible light. Future research may extend this principal and tap into the brain in a more direct fashion, and also transfer information between the senses so that we can see the 'colours' of sounds, smells and tastes. This already happens in the rare condition known as synaesthesia, which mixes sensory modalities. Synaesthesia occurs in many forms and affects from one person in 1,000 to one in 25,000 depending on the region. For example, in colour-grapheme synaesthesia, letters and numbers have innate colours. In some synaesthesic people, A is often perceived as red, B as blue, and C as yellow. In other forms of synaesthesia the person may associate colours with musical notes or phrases, or link tastes and flavours with graphic shapes.

The future: bright and colourful

A journey through vision and colour in the natural world reveals so much that is fascinating and tantalizing. Without eyes to see them, do colours really exist? Light rays of different wavelengths are an undeniable physical phenomenon, but until they are received and acted upon, the concept of colour is not required. Do animals see the world as we do? This is exceptionally unlikely, as many of their eyes work differently to ours. Even for those that have eyes equivalent to our own, the brain's visual processing and mental perception is likely to be quite disparate. Yet there are similarities and shared experiences too. Some form of camouflage is found in almost every group of creatures, and reaction to warning colours is widespread in the animal kingdom.

Will we ever know the colours of long-gone animals, plants and their habitats from millions of years ago? The technology to do so has advanced enormously in the past few decades, and gradually those prehistoric worlds are receiving more and more colourful makeovers. Moving to modern times, what do colours mean to us? Each individual has favourites, dislikes, associations and reactions that are 'coloured' by personal experiences.

Doubtless nature's uses of myriad colours, and the eyes and visual systems of innumerable creatures, will demand further research, inspire technology, and encourage art and leisure into the future.

Index

Further Information

BOOKS

Cronin, T. W. and Johnsen, S., *Visual Ecology*, Princeton University Press, 2014.

Glaeser, G. and Paulus, H. F., *The Evolution of the Eye*, Springer, 2015.

Holland, M., *Animal Eyes*, Arbordale Publishing, 2015.

Land, M. F. and Nilsson, D. E., *Animal Eyes*, Oxford Animal Biology Series, 2012.

Lazareva, O. F., Shimizu, T. and Wasserman, E. A., *How Animals See the World: Comparative Behavior, Biology, and Evolution of Vision*, Oxford University Press, 2012.

Levin, L. A., Nilsson, S. F. E., et al., *Adler's Physiology of the Eye*, Saunders, 2011.

Parker, A., *In the Blink of an Eye: How Vision Kick-started the Big Bang of Evolution*, Natural History Museum, 2016.

Parker, A., *Seven Deadly Colours: The Genius of Nature's Palette*, Natural History Museum, 2016.

Schwab, I. R., *Evolution's Witness: How eyes evolved*, Oxford University Press, 2012.

WEBSITES

Please note: all website addresses are subject to change.

National Center for Biotechnology Information
www.ncbi.nlm.nih.gov/pmc/articles/PMC2781854/

Nautilus
http://nautil.us/issue/11/light/how-animals-see-the-world

Prezi
https://prezi.com/4coymf0pkmvh/evolution-of-the-eye/

The Scientist online
www.the-scientist.com/?articles.view/articleNo/41055/title/The-Rainbow-Connection/

Wikipedia
https://en.wikipedia.org/wiki/Evolution_of_the_eye

Picture credits